RESTORING PRAIRIE WETLANDS

PRAIRIE

A Special Publication of Ducks Unlimited's

RESTORING
WETLANDS

An
Ecological
Approach

**Susan M. Galatowitsch
and Arnold G. van der Valk**

Institute for Wetland and Waterfowl Research

IOWA STATE UNIVERSITY PRESS / AMES

Susan M. Galatowitsch is an assistant professor in the Department of Horticultural Science and the Department of Landscape Architecture, University of Minnesota, where she teaches restoration ecology and is doing wetland research. She received her doctorate from Iowa State University, where she researched wetland restorations in the southern prairie pothole region.

Arnold G. van der Valk is a professor in the Department of Botany, Iowa State University, where he has been teaching wetland ecology and doing wetland research for over 20 years. His research projects on wetland vegetation dynamics, wetland nutrient cycling, and wetland restoration and creation have taken him all over the world.

The Institute for Wetland and Waterfowl Research (IWWR) was established in 1991 by the boards of directors of Ducks Unlimited (DU) in the United States, Canada, and Mexico. DU is a nonprofit wetlands conservation organization whose mission is to fulfill the annual life cycle needs of waterfowl throughout North America. One of IWWR's objectives, pursued through this publication, is to enhance the communication of the latest information on wetlands and waterfowl biology and conservation. Information contained in this publication provides a comprehensive and technical basis for the restoration of wetlands in the southern prairie pothole region of the United States, among the most important and seriously impacted habitats for waterfowl and other wildlife in North America.

Further information on IWWR and its programs can be obtained by writing to IWWR, c/o Ducks Unlimited, Inc., One Waterfowl Way, Memphis, TN 38120-2351.

Library of Congress Cataloging-in-Publication Data
Galatowitsch, Susan M.
    Restoring prairie wetlands: an ecological approach / Susan M. Galatowitsch and Arnold van der Valk.–
1st ed.
        p.   cm.
    Includes bibliographical references and index.
    ISBN 0-8138-2499-0 (alk. paper)
    1. Wetland conservation. 2. Wetland conservation–Prairie Pothole Region. 3. Restoration ecology. 4. Restoration ecology–Prairie Pothole Region. 5. Wetland ecology. 6. Wetland ecology–Prairie Pothole Region. I. Valk, Arnold van der. II. Title.
QH75.G35   1994
333.91'8153'0977–dc20   93-27825

CONTENTS

PREFACE

A farmer whose youth was spent
laying tile lines and digging ditches is now breaking these tiles and filling
these ditches. Marshes nearly exterminated by drainage are reappearing
on his farm. This farmer is not unique. Agriculturists who built careers
planning farm drainage now devise wetland restoration plans, while
wildlife biologists who spent their lifetimes watching wetlands being
drained and waterfowl populations declining now have reason to be
optimistic. Through the collective efforts of farmers, conservationists,
government agents, and wetland and wildlife ecologists, thousands of
wetlands have been restored in the prairie pothole region since 1985.

The restoration of wetlands indicates a significant change in the
economic, social, and political forces that have shaped the prairie landscape for the last 100 years. Not only have economic and political reali-
ties changed but a new factor, the environment, has become increasingly
important. The negative consequences of converting as much land as
possible, regardless of its suitability, into cropland are being recognized.
These include the eradication of most of the prairies, forests, and wetlands; the alteration of our regional hydrology with larger volumes of
water entering our streams and rivers more quickly; and the contamination of surface and groundwater by nutrients and pesticides in runoff
from agricultural fields. One of most important features of the preagri-
cultural prairie landscape was its wetlands. They provided habitat for an
astonishing variety of plants and animals, provided local water storage
capacity, and improved water quality by removing sediment and nutri-

ents from water passing through wetlands. Our hope is that restored wetlands will function like natural wetlands and that, as a consequence, environmental conditions in the prairie pothole region will improve significantly for all of the region's inhabitants. The future of wetland restoration programs is ensured, however, only if restored wetlands actually provide the habitat, water quality, and water storage benefits that supporters of restoration programs promise.

The basic design and construction specifications for wetland restorations are well established and have not changed much in the past 25 years. Nevertheless, how closely restored wetlands resemble natural wetlands in composition and function has been little investigated. Most of the restoration effort in the prairie pothole region is based on the untested assumption that removing a drainage tile or blocking a drainage ditch will inevitably result in the restoration of a wetland. Is this assumption valid? Is there nothing to be learned from the thousands of restorations that have been done that will improve the chances of success of future restorations? We think a careful examination of restoration success and failure is essential for improving site selection criteria and design of restorations and for evaluating the functioning of restored wetlands. It is the responsibility of those entrusted with the restoration of wetlands to see to it that wetland restoration projects are designed and carried out so that restored wetlands actually have the features and functions of natural wetlands.

This book links what is known about wetland ecology to the techniques commonly used to restore prairie wetlands. To our knowledge, this has not been previously attempted. Books and manuals exist that describe wetland restoration techniques, but none were meant to help private landowners and people working for conservation organizations and agencies to develop restoration plans and to improve their projects after construction. In the same way that county soil surveys and state drainage guides need to be specific to an area to be useful in the field, this book focuses on a specific region. This region is that part of the prairie pothole region where row crop agriculture covers more than one-half of the land (Figure i.1). It also happens to be an area where restoration efforts have been particularly intensive and are likely to continue on a large scale in the future.

It is hoped that the approach described here in detail for the southern prairie area will be a useful model for areas beyond its geographic range. The information contained here should not be extrapolated to places where it does not apply. Nor does it seem useful to approach wetland restorations in a way that is so general as to ignore the distinctiveness of the various wetlands in different regions of North America,

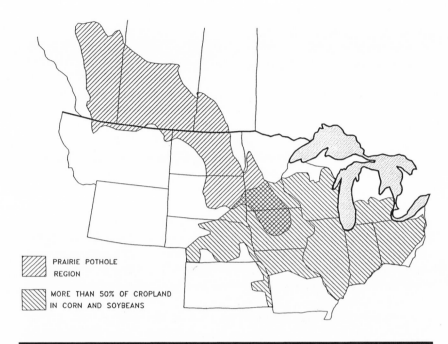

Figure i.1. The southern prairie pothole region is the part of the continent where the prairie pothole region overlaps with the corn belt. More than 50% of cropland is in corn and soybeans. (Map partly based on one in Onstad et al., 1991.)

especially important geographic differences of plant, animal, and soil distributions, changes in hydrology resulting from different landforms, and variations in land use. Rather, this approach, if successful, could be a template for developing comparable manuals for other regions.

Individuals planning to restore wetlands in other portions of the prairie pothole region will find the revegetation and wildlife information largely relevant. For the northern part of the prairie pothole region, plant and animal information in this book would need to be expanded to add species that are found in alkaline and saline wetlands. Soils and hydrology, however, differ significantly between southern and northern portions because of considerably lower precipitation in the north and west. Soil lists and hydrologic information will need to be compiled for the northern area. The basic site selection and design techniques are relevant for wetland restorations of most depressional basins in the mid-continent area, not just the southern prairie pothole region.

Research for this book included several years of field work by Susan

Galatowitsch. Ducks Unlimited, Inc., the North American Wildlife
Foundation, the National Fish and Wildlife Foundation, and the Na-
tional Audubon Society supported various aspects of her field research.
Staff from the U.S. Fish and Wildlife Service, U.S. Soil Conservation
Service, Minnesota Board of Water and Soil Resources, Minnesota Department of Natural Resources, and the Iowa Department of Natural
Resources provided project information and helped with field logistics.
Many landowners permitted access to their property and provided helpful insights. Craig Erickson illustrated the construction figures in Chapter 5 and provided technical support throughout the production of this
book. We would like to thank the following individuals who reviewed
portions of the book: James Baker (Iowa State University, Department
of Agricultural Engineering), William Crumpton (Iowa State University,
Department of Botany), James Dinsmore (Iowa State University, Department of Animal Ecology), Lisa Hemesath (Iowa Department of Natural Resources, Nongame Program), Robert Hoffman (Ducks Unlimited, Inc.), Daniel Mason (Wetlands Research, Inc.), Robert Meeks
(Ducks Unlimited, Inc.), Michael Thompson (Iowa State University, Department of Agronomy), and two anonymous reviewers.

We would particularly like to thank R. A. N. Bonnycastle of Cal-
gary, Alberta, Canada, who generously provided funds to help with the
publication of this manual through a contribution to the Institute for
Wetland and Waterfowl Research (IWWR). IWWR, which was established in 1991, is the research branch of Ducks Unlimited, Inc.; Ducks
Unlimited Canada; and Ducks Unlimited Mexico. One of IWWR's objectives is to make available the latest information on wetland and waterfowl biology in order to enhance the protection, restoration, and management of North American wetlands. We hope that the publication of
this book in some small measure contributes to IWWR's mission.

We dedicate this book to Bruce Batt and the staff of the IWWR.
Bruce was a strong supporter of this project from its inception and he
continued to support it as he migrated from Delta to Ducks Unlimited to
IWWR, from Canada to the United States, and from Chicago to Memphis.

REFERENCE

Onstad, C.A., M.R. Burkhart, and G.D. Bubenzer. 1991. Agricultural research
to improve water quality. Journal of Soil and Water Conservation 46(1):
184–188.

RESTORING PRAIRIE WETLANDS

INTRODUCTION

By the 1930s, the Izaak Walton League and other conservation groups alarmed by declining waterfowl populations called for the replacement of the 84 million acres of wetland habitat lost by drainage in the United States (Page et al., 1938). The first major program designed to restore wildlife habitat was the Federal Aid in Wildlife Restoration Act (Pittman-Robertson Act) of 1937 that provided much of the funding necessary to implement wildlife management practices (Suring and Knighton, 1985). Among the first habitat restorations attempted were the reflooding of peatlands that had been marginal cropland and prone to fire since agricultural drainage. By the mid-1940s small marshes were being constructed throughout much of the northeastern United States.

Waterfowl censuses in 1947 and 1948 indicated that the long downward trend in the continental waterfowl population had been stopped. To what extent marsh construction and restoration or other management practices influenced waterfowl populations is unknown, but the possibility that creating more wetland habitat might increase duck populations spurred an interest in wetland creation and restoration. Wildlife managers from the northeastern United States called for the construction of small-scale water impoundments throughout "the northern States and southern Canada, west to the Plains States and Prairie Provinces" (Mendall, 1949). The impoundments that were subsequently constructed were mostly on publicly owned wildlife management areas or private hunting areas. The first criteria developed to evaluate potential impoundment sites were published in 1951 (Bradley and Cook, 1951). Proper site selection and engineering plans were recognized as critical for new impound-

3

ments to develop into productive waterfowl habitat. Subsequent studies of these impoundments and restorations were conducted in the 1950s and 1960s and showed the need for water-level management to sustain desirable vegetation types (Uhler, 1956; Harris and Marshall, 1963; Kadlec, 1960, 1962).

In the mid-1980s, the implementation of the North American Waterfowl Management Plan, Conservation Reserve Program (CRP) of the 1985 Food Security Act (Farm Bill), and state legislation such as Reinvest in Minnesota created unprecedented opportunities to restore wetlands in the prairie pothole region, particularly on private lands. In the first four years of full implementation of the CRP, more than 1,800 basins (Figure 1.1) were restored on CRP land in the southern prairie pothole region (Galatowitsch, 1993). These restorations were done by either disrupting drainage tiles or plugging drainage ditches. After their restoration, these wetlands have generally not been monitored to determine what kind of wetland developed or even if a wetland developed at all. The widely held assumption is that removing or plugging the drainage system of a drained basin will reestablish its former hydrology. The other widely held assumption is that restoring a basin's former hydrology will result in the natural revegetation of the basin and its recolonization by various groups of invertebrate and vertebrate animals. Field data and observations suggest that these two assumptions may not necessarily always be valid.

The landscapes in which wetland restorations are being done in the prairie pothole region are significantly different from presettlement landscapes in many ways. Regional groundwater levels have been lowered by drainage. In fact, difficulties restoring semi-permanent and permanent wetlands that formerly had groundwater inputs have already been observed in northern Iowa (Galatowitsch, 1993). Seed banks of drained wetlands have been decimated by years of cultivation and little or no input of new wetland seed. There are very few natural wetlands remaining that can act as sources of the seeds, spores, and other propagules of species not present in restored wetlands. Restored wetlands are more isolated because there are relatively few of them and they are not normally part of a wetland complex as were most presettlement wetlands. Thus, it is likely that many restored wetlands may not resemble in the foreseeable future those that formerly occurred in their basins in either composition or function. In short, whether or not it is possible to restore wetlands similar to those that existed historically is uncertain. This does not imply that restored wetlands will necessarily be inferior to wetlands that existed historically in their basins, just that they may be different to some unknown degree in composition and possible function. The chal-

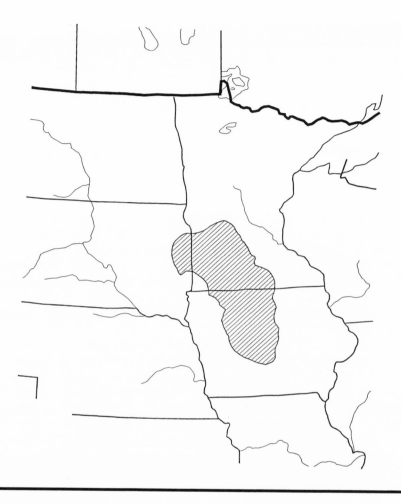

Figure 1.1. In the southern prairie pothole region, 1892 restorations were completed between 1987 and 1991.

lenge of wetland restoration in the prairie pothole region is to recreate wetlands comparable to those that existed over a century ago, but within today's agricultural landscapes.

The revegetation of restored wetlands is assumed to be rapid because of the presence of a relict seed bank that contains seeds of wetland species that formerly grew there. If this were so, the revegetation of restored basins would only require creating suitable environmental conditions for seed germination. However, it has been shown that the seed banks of tile-drained wetlands are often depauperate, and usually dominated by seeds of agricultural weeds and of a few wind-dispersed species, such as cattails (Wienhold and van der Valk, 1989). Only basins drained and cultivated less than 20 years have seed banks that resemble those in natural wetlands, and only these recently drained basins will become revegetated as a result of recruitment from their seed banks when they are restored (Galatowitsch, 1993). (Basins drained with ditches often are revegetated by plant populations that persisted in the ditches and in parts of the basins that were only partially drained.) Revegetation of tile-drained restored wetlands will occur only if propagules of wetland species, especially those of submersed aquatics and sedge meadow species, reach them. Most of these species are water or animal dispersed. Restored prairie potholes are in a landscape that has only a few natural wetlands remaining. In the short term, dispersal between restored wetlands and natural wetlands may not be reliable enough to revegetate them. Transplanting, using donor seed bank material from a suitable natural wetland, or sowing seeds of species with poor dispersal potentials may be necessary because lag times associated with natural dispersal are unacceptably long.

Twenty to 32 families of invertebrates are typically found in natural wetlands in Iowa (Voigts, 1976) whereas only 4 to 16 families per site are found in restored wetlands (LaGrange and Dinsmore, 1989; Hemesath, 1991). It is possible that dispersal problems, comparable to those for plant species, may also slow down or even prevent the recolonization of restored wetlands by some groups of invertebrates.

Poor revegetation and invertebrate recolonization in restored prairie potholes has consequences for the reestablishment of wildlife. Twenty-two (Table 1.1) of the 54 breeding bird species of the region (as listed in Appendix B) have been observed to nest in restored wetlands (Sewell, 1989; Delphey, 1991; Hemesath, 1991). Some of these species, however, occur in lower abundances in restorations without well-developed vegetation, including sora (*Porzana carolina*), Virginia rail (*Rallus limicola*), marsh wren (*Cistothorus palustris*), common yellowthroat (*Geothlypis trichas*), yellow-headed blackbird (*Xanthocephalus xanthocephalus*), red-winged blackbird (*Agelaius phoeniceus*), and swamp sparrow (*Melospiza georgiana*). Amphibians, particularly tiger salamanders (*Ambystoma tigrinum*), have readily recolonized newly restored wetlands. Amphibian surveys of the restored McBreen marshes in Dickinson County, Iowa, found many tiger salamander and leopard frog tadpoles (*Rana pipiens*), but tadpoles of both species were 75–100% larger than is typical in natural wetlands. Because tadpoles eating each

Table 1.1. Birds observed to breed in restored wetlands within the region

American bittern	Northern shoveler
American coot	Pied-billed grebe
Blue-winged teal	Red-winged blackbird
Canada goose	Sedge wren
Common yellowthroat	Sora
Gadwall	Spotted sandpiper
Green-winged teal	Swamp sparrow
Least bittern	Virginia rail
Mallard	Wilson's phalarope
Marsh wren	Wood duck
Northern pintail	Yellow-headed blackbird

Source: Information from Sewell, 1989; Hemesath, 1991; LaGrange, 1985; and Delphey, 1991.

other are much larger than tadpoles eating invertebrates, the tadpoles in these restored wetlands evidently had become cannibalistic. In fact, invertebrate populations were extremely low in these wetlands, and the amphibians seem to have reached them before an invertebrate food base had become established (Lannoo et al., 1993).

The ongoing restorations in the southern prairie pothole region may represent the first opportunity to reestablish a major waterfowl breeding area within North America. However, it remains to be seen if the new habitat created will be sufficient to support ducks that historically bred in the region. The mallard (*Anas platyrhynchos*) and blue-winged teal (*Anas discors*) are, by far, the most common species on restored wetlands. Sewell (1989) and Delphey (1991) both reported approximately 75% of all pair counts as either of these species. Likewise, Zenner, Grange, and White (1990) observed average breeding pairs/square mile of water to be 6.8 for the mallard, 3.5 for the blue-winged teal, and 1.3 for all other ducks. Conditions may not yet be suitable for other species, such as the northern pintail (*Anas acuta*) and gadwall (*Anas stepera*). Since gadwalls and pintails are sensitive to landscape-level habitat quality, such as amount of uncultivated land and number of ephemeral ponds, restored wetlands per se do not necessarily offer suitable habitat for these species. In addition, predation of waterfowl using restored wetlands needs to be investigated.

The examples cited suggest that a restored wetland with only some characteristics of a natural wetland, e.g., a stand of cattails or a pair of blue-winged teal, should not be considered comparable to a natural wetland in function and that such restorations should not be classified as successes. Currently wetland restoration success is measured only in terms of the number of basins restored. This is both overly simplistic and highly misleading. Simply quantifying the number of basins as a measure of success elevates any depression with water, and even some without water, to the status of a wetland. The success of a wetland restoration can only be determined by comparing its characteristics to those of similar natural wetlands. Initially, measures of restoration success will have to be based on similarities in species composition between restored and natural wetlands. Ideally, they will eventually be based on comparisons of the functioning of restored and natural wetlands, e.g., their primary and secondary production and rates of denitrification.

We believe that restoring prairie wetlands so that they have the characteristics of natural wetlands may require more than simply breaking a tile or plugging a ditch. We do not deny, however, that there are situations where this will be adequate. What we are suggesting is that not all pothole restorations can be done in the same way. There are significant differences in the duration of drainage of basins, how the basin was drained, farming activities in the drained basin, types of soil in the basin, changes in the effective size of the basin's watershed, surrounding land use, etc., that should be taken into account in planning and carrying out a restoration. Taking these factors into account will improve the chances of success for future restorations because it will result in better site selection, project design, and project construction. Even problems with existing restorations can be corrected.

This book examines basic ecological, hydrological, and technical considerations in planning, constructing, managing, and evaluating wetland restorations in the southern prairie pothole region, i.e., those parts of the prairie pothole region in Iowa, South Dakota, and Minnesota (Figure 1.2). The information and recommendations in it are based on the scientific literature, our own fieldwork on and visits to restored wetlands all over the southern prairie pothole district, and many productive conversations and meetings with those carrying out these restorations for various state and federal agencies and conservation organizations. Five major topics are covered: (1) the basic ecology of prairie potholes and how prairie wetlands have changed since European settlement are reviewed in Chapter 2; (2) guidelines for selecting suitable sites to restore are given in Chapter 3; (3) guidelines for designing restorations that emphasize wildlife habitat and water quality considerations are presented in Chapter 4; (4) the evaluation and management of wetland restorations is considered in Chapters 5 and 6; and (5) landowner assistance programs available through government agencies and private conservation groups are described in Chapter 7.

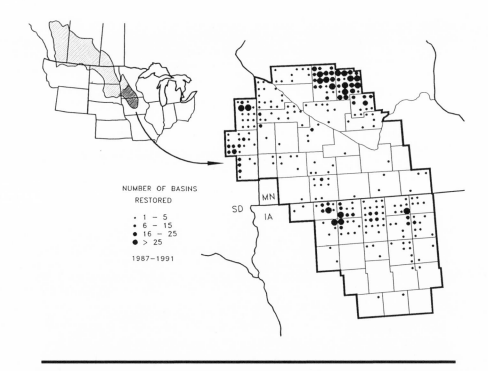

Figure 1.2. The southern prairie pothole region of southeastern South Dakota, southern Minnesota, and northern Iowa.

REFERENCES

Bradley, B.O. and A.H. Cook. 1951. Small marsh development in New York. Transactions of the North American Wildlife Conference 16: 251–264.

Delphey, P.J. 1991. A comparison of the bird and aquatic macroinvertebrate communities between restored and natural Iowa prairie wetlands. M.S. Thesis, Iowa State University, Ames.

Galatowitsch, S.M. 1993. Site selection, design criteria, and performance assessment for wetland restorations in the prairie pothole region. Ph.D. Dissertation, Iowa State University, Ames.

Harris, S.W. and W.H. Marshall. 1963. Ecology of water-level manipulations on a northern marsh. Ecology 44: 331–343.

Hemesath, L.M. 1991. Species richness and nest productivity of marsh birds on restored prairie potholes in northern Iowa. M.S. Thesis, Iowa State University, Ames.

Kadlec, J.A. 1960. The effect of a drawdown on the ecology of a waterfowl impoundment. Game Division Report 2276. Michigan Department of Conservation, Lansing. 181 pp.

Kadlec, J.A. 1962. Effects of a drawdown on a waterfowl impoundment. Ecology 43: 267–281.

LaGrange, T.G. 1985. Habitat use and nutrient reserves dynamics of spring migratory mallards in central Iowa. M.S. Thesis, Iowa State University, Ames.

LaGrange, T.G. and J.J. Dinsmore. 1989. Plant and animal community responses to restored Iowa wetlands. Prairie Naturalist 21: 39–48.

Lannoo, M.J., K. Lang, T. Waltz, and G.S. Phillips. 1993. Profile of an abraded amphibian assemblage: Dickinson County, Iowa, 70 years after Frank Blanchard's survey. Unpublished manuscript.

Mendall, H.L. 1949. Breeding ground improvements for waterfowl in Maine. Transactions of the North American Wildlife Conference 14: 58–64.

Page, J.C., S.B. Locke, S.H. McCrory, and G.W. Grebe. 1938. What is or should be the status of wildlife as a factor in drainage and reclamation planning? Transactions of the North American Wildlife Conference 3: 109–125.

Sewell, R.W. 1989. Floral and faunal colonization of restored wetlands in west-central Minnesota and northeastern South Dakota. M.S. Thesis, South Dakota State University, Brookings. 46 pp.

Suring, L.H. and M.D. Knighton. 1985. History of water impoundments in wildlife management. IN M.D. Knighton (Ed.). *Water Impoundments for Wildlife: A Habitat Management Workshop.* U.S. Forest Service General Technical Report NC-100. North Central Forest Experiment Station, St. Paul, Minnesota.

Uhler, F.M. 1956. New habitats for waterfowl. Transactions of the North American Wildlife Conference 21: 453–469.

Voigts, D.K. 1976. Aquatic invertebrate abundance in relation to changing marsh vegetation. American Midland Naturalist 95: 313–322.

Wienhold, C.E. and A.G. van der Valk. 1989. The impact of duration of drainage on the seed banks of northern prairie wetlands. Canadian Journal of Botany 67: 1878–1884.

Zenner, G.G., T.G. LaGrange, and S. White. 1990. Preliminary results—Iowa Prairie Pothole Joint Venture evaluation: breeding pair estimates and nest success. Unpublished research report. Iowa Department of Natural Resources, Clear Lake. 7 pp.

ECOLOGY
OF
PRAIRIE POTHOLES

PRAIRIE LANDSCAPES

Continental glaciers reached their maximum extent in much of North America 17,000 to 21,000 years ago. As the climate became warmer, these continental glaciers retreated northward, except in what is now southern Minnesota and northern Iowa. In spite of climatic warming, a lobe of the continental glacier, now called the Des Moines Lobe, apparently surged southward, spreading a relatively thin mantle of ice into central Iowa 14,000 years ago (Kemmis, 1991). This ice sheet is thought to have stagnated because of the prevailing warm conditions, and it became riddled with a cave-like network of tunnels and rooms. Because this network lacked the debris usually embedded in the rest of the glacial sheet, as the ice melted, the ghosts of these ice caves remained as shallow depressions in a blanket of glacial till. On the Des Moines Lobe, the last glacial advance left wide bands (30–50 miles wide) of hummocky terrain characterized by "linked depression systems," a multitude of saucer-like depressions connected by shallow drainage ways. These poorly drained and frequently flooded depressions are the potholes. They are called "prairie potholes" because the surrounding upland natural vegetation was prairie at the time of European settlement.

Not all of the landscapes of the prairie pothole region were affected by surging glaciers and stagnant ice sheets. In fact, the Des Moines Lobe is distinctive in this regard. The prairie pothole region north of the Des Moines Lobe was molded by more "typical" glacial conditions and consequently has fewer surface links between potholes. But its distinctive glacial topography is not the only distinguishing feature of the southern

11

prairie pothole region. It is also warmer and wetter than the rest of the region (Figure 1.2). Consequently, conditions were favorable for row-crop agriculture. Even though the Des Moines Lobe was considered inhospitable, if not uninhabitable, by early European explorers, an interest in draining its wetlands quickly developed as an appreciation of its potential as farmland grew.

Wetland drainage started before 1900 and began in earnest after the turn of the century. Drainage wells, ditches, and later tile lines were used to drain areas that were in presettlement days up to 70% wetland. By 1925, less than 1% of the land in north-central Iowa was classified as "rough swamp land" (Iowa State Census Reports, 1925), and some negative effects of drainage had become apparent. Flowing wells used as water sources, that had been sunk from 10 to 60 feet, went dry, and wildlife populations declined (Flickinger, 1904; Birdsall, 1915).

Prairie Potholes

Before settlement, the ridges, knobs, and other rises once supported tallgrass prairie: big bluestem (*Andropogon geradii*), little bluestem (*Andropogon scoparius*), indiangrass (*Sorghastrum nutans*), and a myriad of prairie flowers and other prairie grasses. On the highest ridges grew the prairie plants most tolerant of dry conditions while in the intervening depressions and swales grew wetland plants able to withstand at least periodic ponding. The transition from ridge to wetland vegetation was gradual, with drier prairie plants replaced by wet prairie plants on toe slopes. Wet prairie included a few plants from the ridges — such as switchgrass (*Panicum virgatum*) along with a new suite of species including prairie cord grass (*Spartina pectinata*). In the depressions that contained standing water at least seasonally during wet years, wetland plants replaced wet prairie species. Sedge meadow perennials occupied areas whose soil was saturated with water into early summer while emergent species occupied areas with prolonged periods of standing water. In areas where there was standing water throughout most summers, submersed and floating leaved aquatics were the dominant species. Thus, concentric rings or zones of vegetation occur in these wetlands because groups of plant species with similar water depth or flooding tolerance are found growing together. These vegetation zones are usually very distinct because the dominant species in each zone have different growth forms.

Nearly 350 species of plants are found in the wetlands of the southern prairie pothole region (Appendix A1). Only one-sixth to one-third of the 350 species will be found in any one wetland. Whether a particular species of plant is found in a wetland will depend to some extent on the vagaries of seed dispersal and the suitability of conditions for its establishment. Existing vegetation can prevent seeds of species that have reached a wetland from germinating. Most wetland species are perennials capable of clonal growth that can spread rapidly. It is only after a disturbance removes some of the existing vegetation that seed germination typically occurs for most species.

Stewart and Kantrud (1971) developed a classification of prairie potholes based on the vegetation found in the deepest zone of wetland basins. There are five different vegetation types in freshwater prairie wetlands: wet prairie, sedge meadow, shallow marsh, deep marsh, and permanent open water. The maximum depth and duration of standing water will determine which vegetation type will occupy the deepest area of a basin (Figure 2.1). The surrounding, shallower parts of the basin will be occupied by vegetation less tolerant of flooding. So five wetland classes can be distinguished: Class I, ephemeral ponds with a central wet prairie zone, Class II, temporary ponds with a central sedge meadow zone, Class III, seasonal ponds and lakes with a central shallow marsh zone dominated by bulrushes and other emergent marsh plants, Class IV, semi-permanent ponds and lakes with a central deep marsh zone, and Class V, permanent ponds and lakes with a central permanent open water zone. In addition, fens are a special kind of sedge meadow with peaty soils high in carbonates that are found in areas where groundwater discharges, i.e., where there are springs and seeps. Fens can occur in flat areas but often are found on the side slopes of basins.

The Stewart and Kantrud (1971) system for classifying prairie potholes predates the development of a national wetland classification system (Cowardin et al., 1979). The Stewart and Kantrud system classifies entire basins while the ultimate field units classified by the Cowardin et al. system are floristically homogeneous stands of vegetation. Thus the vegetation in a pothole using the Cowardin et al. system would normally be placed into a number of different vegetation classes. For additional information on the application of both systems to prairie potholes see Kantrud et al. (1989). Within a restoration context, the Stewart and Kantrud system is more useful than the Cowardin et al. system because it does classify entire basins, and it is entire basins that are being restored. We have used the Stewart and Kantrud system to classify natural and restored potholes throughout this book. What follows is a brief description of each class and the common species in its defining vegetation type.

Wet prairies (Class I) are usually dominated by grasses. Common species include bluegrass (*Poa* spp.), big bluestem, and switchgrass. Other important species include death camas (*Zigadenus elegans*), Mich-

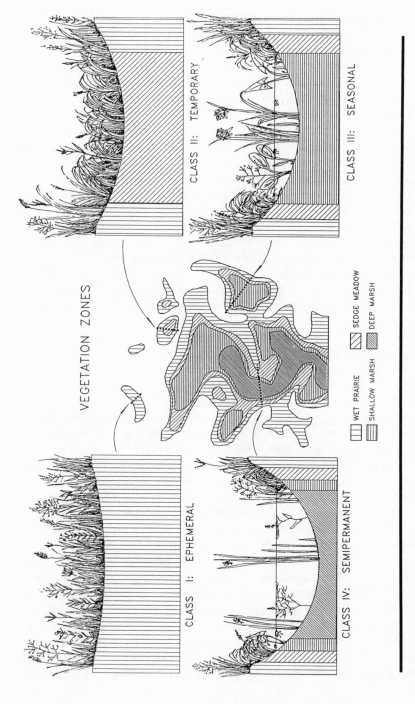

Figure 2.1. Wetland classes and their associated wetland zones. (Adapted from Stewart and Kantrud, 1971)

igan lily (*Lilium philadelphicum*), and meadow parsnip (*Zizia* spp.).

Sedge meadows (Class II) have a periphery of wet prairie and a central area generally dominated by several of many possible sedges such as woolly sedge (*Carex lanuginosa*), clustered field sedge (*Carex praegracilis*) and green bulrush (*Scirpus atrovirens*) along with grasses such as bluejoint (*Calamagrostis canadensis*) and prairie cord grass. Other common species include Torrey's rush (*Juncus torreyi*), Dudley rush (*Juncus dudleyi*), woundwort (*Stachys palustris*), and bugleweed (*Lycopus* spp.).

Seasonal marshes (Class III) have a central shallow marsh zone surrounded by sedge meadow and wet prairie. The dominant species in the shallow marsh zone are coarse sedges, some grasses, and midheight herbaceous species including river bulrush (*Scirpus fluviatilis*) and broad leaf cattail (*Typha latifolia*). Awned sedge (*Carex atherodes*) and lake sedge (*Carex lacustris*) tolerate frequent drying and are very common species in shallow seasonal wetlands. Also common is beaked sedge (*Carex rostrata* var. *utriculata*), which needs continually wet conditions. Other typical species include common burreed (*Sparganium eurycarpum*), water plantain (*Alisma triviale*), and water smartweed (*Polygonum amphibium*). Open water pockets have floating and submersed species such as slender riccia (*Riccia fluitans*), star duckweed (*Lemna trisulca*), common bladderwort (*Utricularia vulgaris*), and leafy pondweed (*Potamogeton foliosus*).

Semi-permanent (Class IV) marshes are surrounded by the previously described vegetation types and have a deep marsh zone that is dominated by hard-stemmed bulrush (*Scirpus acutus*), soft-stemmed bulrush (*Scirpus validus*), river bulrush, and hybrid cattail (*Typha glauca*). Semi-permanent wetlands have similar open water species as the preceding. During periods of drought, deep water areas can become completely dry, exposing bare mud. These mudflats are dominated by annual plants, such as smartweeds (*Polygonum pensylvanicum* and *Polygonum lapathifolium*), beggar's ticks (*Bidens cernua*), and nutsedges (*Cyperus* spp.).

Landscape Patterns

Just as the vegetation in the deepest part of the basin reflects the hydrology of a wetland basin, basin hydrology can also be inferred from its soils. The degree of saturation and movement of water through soils as they formed in glacial deposits resulted in noticeable variation in soil characteristics such as color, organic matter content, and particle size distribution. Many of these characteristics persist long after artificial drainage, leaving a historic record of presettlement vegetation and hydrology. For example, a vertical slice of soil (a profile) in a semi-permanent pond (Class IV) will reveal a thick layer (horizon) of partially decomposed plant material and snail shells (sometimes several feet thick), whereas a wet prairie (Class I) has a thinner, more decomposed organic layer covering a gray mineral layer. The gray layer shows that the soil had been saturated and consequently the iron oxides (rust) are not as abundant as in upland soils.

County soil surveys originally made to guide agricultural use can be used to estimate wetland extent before drainage and cultivation (Hewes, 1951). Soil maps suggest that at the time of settlement, wetlands covered 20% to 60% of northern and central Iowa, the extent varying with landform (Galatowitsch, 1993). Where rivers dissect an area, wetland acreage was low because of the natural drainage of these landscapes. In contrast, very flat areas that lacked natural drainage were predominantly wetland. Hilly areas formed by moraines were roughly 30%–50% wetlands (Figure 2.2).

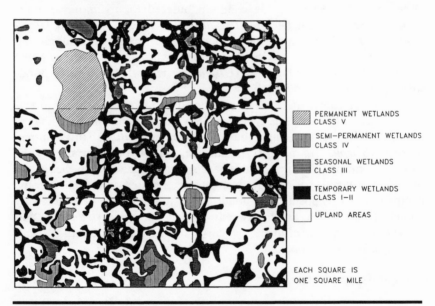

Figure 2.2. The predrainage extent of wetlands in a 9-square-mile area of Wright County (parts of T92N R24W, T92N R25W, T93N R24W, T93N R25W), Iowa, as estimated from county soil maps. Wetland classes correspond to soil map units as follows:

WETLAND CLASS	SOIL MAP UNITS
CLASS V	Open water
CLASS IV	Palms (211), Okoboji muck (90)
CLASS III	Okoboji silty clay loam (6)
CLASS I–II	Webster (107), Canisteo (507), Harps (95), Kossuth (388)

Occasional lakes, most still undrained, occur throughout the region, generally occupying between 1% and 10% of a township (36 square miles) in northern and central Iowa (Galatowitsch, 1993). Lakes are more extensive in other morainal areas at the northern edge of the region (e.g., Kandiyohi County, Minnesota). Semi-permanent (Class IV) and seasonal wetlands (Class III) were more numerous and extensive before agricultural drainage, together occupying roughly four times the acreage of the lakes. These semi-permanent wetlands were usually much larger than seasonal wetlands, but seasonal wetlands were more numerous. Approximately equal acreages of semi-permanent and seasonal wetlands occurred — 700 to 1,000 acres of each type in a township. For every acre of semi-permanent or seasonal wetlands, 4 to 9 acres of wet prairie and sedge meadow occurred around the periphery and along connecting swales and flats. Thus, the presettlement prairie landscape typically was dotted with many small temporary and seasonal ponds interspersed with fewer, larger, semi-permanent or permanent basins. Most townships (36 square miles) had between 4,000 and 12,000 acres of different wetland types prior to European settlement.

Vegetation Dynamics

Seeds reaching or produced in a wetland when conditions are unfavorable for establishment may remain viable in the sediments for years. Some seeds, especially those of mudflat annuals, such as smartweeds, and emergents, such as bulrushes, can often survive for decades whereas seeds from sedges and submersed aquatics may survive only a few years or less (van der Valk and Davis, 1979). Viable seed present in the sediment at any time is referred to as a "seed bank." The seed bank often has a different composition than the vegetation in a wetland, reflecting the composition of past vegetation. It can be used to predict future composition of the vegetation under some circumstances (van der Valk and Davis, 1978).

In response to cyclical changes in water depth caused by years of above-normal annual precipitation alternating with years of below-normal annual precipitation, the vegetation of a semi-permanent or permanent wetland often changes cyclically over a period of 5 to 20 years (Weller and Fredrickson, 1974). During such a cycle (Figure 2.3), emergent vegetation gradually declines after several years of high water and muskrat herbivory (Weller and Spatcher, 1965). Open water conditions persist until the next drought lowers water levels causing all or part of the wetland basin to dry. Severe droughts occur every 10 to 12 years within the region. Annual plants and emergents recruited from the seed bank quickly colonize exposed mudflats during drawdowns. When these

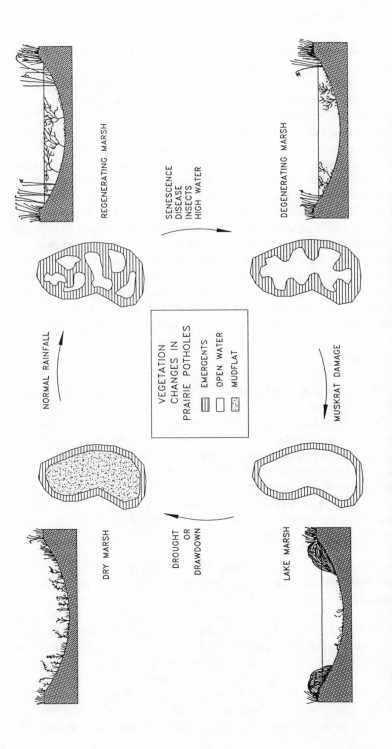

Figure 2.3. Vegetation dynamics of prairie pothole wetlands. (Adapted from van der Valk and Davis, 1978)

REGENERATING MARSH

SENESCENCE
DISEASE
INSECTS
HIGH WATER

DEGENERATING MARSH

NORMAL RAINFALL

VEGETATION
CHANGES IN
PRAIRIE POTHOLES

EMERGENTS
OPEN WATER
MUDFLAT

MUSKRAT DAMAGE

DRY MARSH

DROUGHT
OR
DRAWDOWN

LAKE MARSH

wetlands reflood as the drought ends, annuals are eliminated, and emergent plants spread vegetatively over the basin, starting another cycle.

Shallow areas of semi-permanent marshes and shallower classes of potholes often do not have pronounced vegetation cycles (van der Valk and Davis, 1979). Evaporative losses are sufficient to cause shallow wetlands to dry annually, rather than only periodically. Consequently, shallow marshes are continually occupied by plants that are adapted to ephemeral and seasonal ponding, such as sedges. In contrast, deep water marsh areas oscillate between submersed aquatics and annuals that grow only on mudflats. Besides climatically controlled vegetation changes, smaller scale disturbances, such as those created by herbivorous animals can also create opportunities for new plants to become established within wetlands from the seed bank.

As a result of studying changes in waterfowl use of wetlands during vegetation cycles, wetlands with a 50:50 cover of vegetation and open water have long been recognized supporting the maximum number of waterfowl species and numbers (Weller and Spatcher, 1965). And, for nearly 30 years, managing wetlands to sustain this "hemi-marsh" stage has been a goal of wetland managers (Weller and Fredrickson, 1974). Approximately 50% vegetative cover of emergents occurs in wetlands as water levels rise after a drawdown as emergent species spread across the basin and again as the emergent vegetation dies out due to high water or muskrat damage.

The hemi-marsh seems to provide an optimal combination of cover and food for waterfowl. Patches of emergent vegetation interspersed with open water seem to create conditions that support an abundance of nesting sites and invertebrates in hemi-marshes. In fact, dabbling ducks may use this interspersed pattern as a cue to select suitable feeding areas (Nelson and Kadlec, 1984). Rising water levels after periods of drought can kill much of the vegetation, such as annual grasses and smartweeds. The decomposing litter of these plants is used directly as a substrate by invertebrates, and they feed on the algae and other microorganisms growing on it. Studies in Iowa marshes showed that shallow water with standing and floating litter produced the most isopods, planorbid snails, and physid snails, all food for waterfowl (Voigts, 1976).

Wetland Animals

Many factors control the distribution of animals within prairie marshes. In some cases, animals respond to the structure of the vegetation—how well concealed they are from predators, how easily they can travel through an area; other times they may require a particular food source—such as aquatic insects, seeds, or even plant roots. Some wetland animals are generalists and will use a variety of wetland habitats. Tiger salamanders (*Ambystoma tigrinum*), American toads (*Bufo americanus*), painted turtles (*Chrysemys picta*), and snapping turtles (*Chelydra serpentina*) inhabit almost any body of water. Others, such as waterfowl, require a diversity of wetlands to meet changing requirements over their life cycles. Still others require very specific conditions for nesting or feeding. Weller and Spatcher (1965) recognized four categories of marsh-breeding birds based on nest-site selection in central Iowa: (1) birds that select nest sites along the marsh edge in low trees and shrubs, (2) birds that use sedges and grasses for nesting, (3) birds that nest in tall and robust emergent vegetation, (4) birds that use low mats of vegetation, often in open areas. Weller and Fredrickson (1974) observed that concentric zones of vegetation produce a related zonation of birds.

Mammals, reptiles, and amphibians can also be characterized by their affinity for different kinds of wetland habitat. In Appendix B, there is a list of animals found in the southern prairie pothole region. Some animals associated with specific wetlands habitats are described in the following.

Wet prairies have tall, dense grass vegetation; these wetlands thaw earlier in the spring than deeper wetlands and dry earlier in the summer. Ground-nesting birds such as the northern harrier (*Circus cyaneus*), short-eared owl (*Asio flammeus*), savannah sparrow (*Passerculus sandwichensis*), Henslow's sparrow (*Ammodramus henslowii*), swamp sparrow (*Melospiza georgiana*), common yellowthroat (*Geothlypis trichas*), mallard (*Anas platyrhynchos*), and northern shoveler (*Anas clypeata*) nest in wet prairie. Short-tailed shrews (*Blarina brevicauda*) and Franklin's ground squirrels (*Spermophilus franklinii*) are two of the mammals using wet prairies although both use upland areas as well. Short-tailed shrews are likely attracted by the abundance of insects, especially early in the season, in wet prairies. Franklin's ground squirrels feed extensively on wetland plants and animals but may use wet prairies primarily because of their tall, dense cover.

Like wet prairies, sedge meadows thaw early and are an early source of insects and other invertebrates. They stay moist later into the summer, which prolongs insect availability. When large sedges and grasses cover a shallow marsh, they provide a closed canopy but an open understory. The Virginia rail (*Rallus limicola*) and sora (*Porzana carolina*) prefer these kinds of wetlands for nesting and feeding because they can move about more freely than in cattails but still be afforded protection from predatory birds. Some ground-nesting wet prairie birds such as the savannah sparrow, swamp sparrow, common yellowthroat, mallard, and northern shoveler can also nest in sedge meadows, often on the hummocks created by some sedges. Other birds that nest in sedge meadows include the American bittern (*Botaurus lentiginosus*), common snipe (*Capella gallinago*), sedge wren (*Cistothorus platensis*), marsh wren (*Cistothorus palustris*), and LeConte's sparrow (*Ammospiza lecontei*). Some of these species, such as the American bittern, nest on a platform of grasses and rushes constructed on the ground in adjacent wet meadows. Ducks, rails, and bitterns feed in marsh areas adjacent to sedge meadows.

The masked shrews (*Sorex cinereus*) and pygmy shrews (*Microsorex hoyi*) can tolerate a wide range of environmental conditions but seem to prefer sedge meadows because of their reliable invertebrate populations. Two other small mammals that feed on plants and seeds inhabit sedge meadows. Meadow voles (*Microtus pennsylvanicus*) are restricted to sites that are quite moist where they feed on rushes, sedges, and grasses along with insect larvae and fungi. They nest on the ground or on tussocks of grasses and sedges. Meadow jumping mice (*Zapus hudsonius*) feed on invertebrates in the spring and switch to seeds, fruits, and fungi as summer progresses. Chorus frogs (*Pseudoacris triseriata*) occur in a variety of wetland habitats but only breed in areas with lush growth of sedges, grasses, and rushes. Chorus frogs use this vegetation as concealment when they call. Smooth green snakes (*Opheodrys vernalis*), one of few prairie snakes to prefer wetlands, live in sedge meadows feeding on the rich supply of insects this habitat affords (Christiansen and Bailey, 1990).

Shallow emergent marshes provide dense vegetation for concealment as well as prolonged standing water for feeding. They provide suitable habitat for birds that build floating nests or nests anchored on emergent vegetation. The horned grebe (*Podiceps auritus*), eared grebe (*Podiceps nigricollis*), least bittern (*Ixobrychus exilis*), Virginia rail (*Rallus limicola*), sora (*Porzana carolina*), common moorhen (*Gallinula chloropus*), American coot (*Fulica americana*), yellow-headed blackbird (*Xanthocephalus xanthocephalus*), and red-winged blackbird (*Agelaius phoeniceus*) are breeding birds of shallow marshes. Many of these species, such as the least bittern, prefer to nest over standing water in stands of cattail or bulrush. Northern pintails (*Anas acuta*) nest on uplands but require small seasonal wetlands for early brood rearing. Blanding's turtles (*Emyoidea blandingi*) also require uplands for egg laying, specifically sandy areas, but feed in shallow marshes with dense vegetation.

Some species that breed in shallower marshes also breed in deep marshes (or semi-permanent marshes): yellow-headed blackbird, red-winged blackbird, and American coot. In addition, two fish-eating species, western grebe (*Aechmophorus occidentalis*) and pied-billed grebe (*Podilymbus podiceps*), nest in deep emergent marshes. Diving ducks including the canvasback (*Aythya vallisneria*), redhead (*Aythya americana*), and ruddy duck (*Oxyura jamaicensis*) select nest sites over water in semi-permanent marshes. Although cattail and bulrush are preferred nesting cover, large monotypic stands of these species seem to be avoided. Ruddy ducks may use aquatic grasses and sedges for nesting.

One inhabitant of deep emergent marshes, the muskrat (*Ondatra zibethicus*), exerts a great effect on the vegetation dynamics of those wetlands. Muskrats feed nearly exclusively on the bases of emergent plants and use these plants to build lodges. Large populations of muskrats can thrive in semi-permanent marshes for a time and may eventually denude the site of its vegetation, an "eat-out." When that happens, their populations decline precipitously until the marsh regenerates its emergent vegetation, after a drawdown. Muskrats are very prolific and are the prime food source of mink (*Mustela vison*), another common inhabitant of semi-permanent marshes, as well as several other predators. Mink also feed on crayfish, fish, smaller rodents, and waterfowl and their eggs.

Open water areas of permanently flooded marshes provide reliable sources of fish for birds including the great blue heron (*Aredea herodias*), green-backed heron (*Butorides virescens*) (brushy vegetation), black-crowned night heron (*Nycticorax nycticorax*), Franklin's gull (*Larus pipixcan*), Forster's tern (*Sterna forsteri*), and black tern (*Chlidonias niger*). The herons nest in trees (great blue heron) or in shrubs along the water's edge (green-backed heron) or in thick stands of emergent vegetation (black-crowned night heron). Black terns nest on floating vegetation in stands of emergent vegetation; Forster's terns select new, high muskrat lodges or floating mats of vegetation for nesting platforms. Canvasbacks and other diving ducks also frequent open water areas, especially those that are somewhat deep and provide a source of invertebrates and/or wild celery (*Vallisneria americana*) tubers.

Thirty-three shorebird species migrate through the southern prairie pothole region. Shorebirds feed on invertebrates and seeds exposed on mudflats. Some ornithologists believe that continental shorebird populations are limited by the quantity and quality of wintering and migration habitat (Reid et al., 1983). Sheetwater areas in cultivated fields probably provide suitable feeding areas but are likely not as rich a source of invertebrates and seeds as are drawndown wetlands. The killdeer (*Charadrius vociferus*), willet (*Catoptrophorus semipalmatus*), marbled godwit (*Limosa fedoa*), American avocet (*Recurvirostra americana*), and Wilson's phalarope (*Phalaropus tricolor*) are among the migrating shorebirds that are or were breeding birds of the region. Cricket frogs (*Acris crepitans*) also use mudflats to forage for small insects (Christiansen and Bailey, 1991).

Wildlife and Wetland Complexes

Many animals, most notably waterfowl, require wetland clusters or *complexes* composed of a mix of small shallow basins and larger deeper basins. A single wetland, even a large semi-permanent wetland, cannot adequately satisfy all their requirements for food and shelter during different life stages: migration, courtship, brood rearing, and molting (e.g., Swanson and Duebbert, 1989). As waterfowl initiate breeding, rear broods, molt, and migrate, their nutritional requirements shift to meet their changing physiological needs. Seasonal and interannual changes in water levels and temperatures greatly affect the availability of food for migrating and breeding waterfowl. Before agricultural drainage, wetland complexes with their different types of wetlands collectively provided a diverse, reliable habitat for breeding waterfowl. Only isolated wetlands, usually large semi-permanent to permanent marshes, remained after agricultural drainage. These large solitary wetlands generally have fewer kinds of ducks using them compared to wetlands of a similar size within a complex. Consequently, the southern prairie pothole region is no longer a major breeding area for waterfowl but rather is important primarily as a migratory stopover area.

Large deep marshes and even seasonally flooded marshes are usually frozen when the first migrating waterfowl pass through the region. However, the shallow ephemeral basins, sheetwater areas, and temporary wetlands (Classes I and II) thaw much earlier. The abundant invertebrate populations in temporary and seasonal wetlands during the spring and early summer attract breeding pairs (Stewart and Kantrud, 1973). Northern pintails are early migrants along with mallards. Ring-necked ducks and lesser scaup are early spring migrants, arriving in the region before large lakes have thawed and so initially use shallow marshes. LaGrange and Dinsmore (1989) observed the habitat use of migrating mallards in central Iowa. Sheetwater ponds, the modern day remnants of the original temporary prairie wetlands, were used for daytime feeding much more (200 times) than larger permanent wetlands. Birds, however, flew up to 8 miles to spend the night on larger permanent wetlands. Populations of invertebrates increase rapidly in these shallow basins, providing an early season food supply that is unavailable in other, deeper wetlands nearby. Seeds are also available for foraging in these recently thawed areas. Temporary wetlands isolate courting pairs and provide loafing sites for males near nesting hens and can even be used for nesting dabbling ducks in wet years when water remains for more than a few weeks.

During drought periods, even the deepest parts of semi-permanent wetlands are shallow enough to allow foraging by dabbling ducks. And, since seasonal wetlands are dry, semi-permanent wetlands are the principal breeding habitat of dabbling ducks during these dry years. In years of normal to above-average precipitation, higher water levels in semi-permanent wetlands provide over-water nesting cover of emergent vegetation required by diving ducks but restrict dabbling duck feeding and resting to peripheral areas. Semi-permanent wetlands thaw much later than shallow wetlands but the delayed invertebrate availability coincides with the later breeding time of diving ducks (Swanson and Duebbert, 1989). Courtship isolation is thought to be achieved on larger marshes by the visual barriers of an irregular shoreline or tall vegetation surrounding open water areas. Configuration of the shoreline seems to influence the number of breeding pairs a large marsh can accommodate, although little supporting evidence has been published. Hens with broods will use primarily semi-permanent wetlands, and to a lesser extent, seasonal wetlands until the ducklings reach flight stage. Hens will use seasonal wetlands if they are still flooded, moving to more permanent basins as the shallow basins dry. Molting birds cannot fly and so also select large semi-permanent or even permanent wetlands with good stands of emergent vegetation that provide adequate protection from predators.

DRAINAGE AND LAND CONVERSION

Prior to European settlement, groundwater tables in the southern prairie pothole region were very close to the surface and often discharged into depressions as springs and seeps (Figure 2.4). To convert this landscape into agricultural land required lowering water tables. The extensive tile and ditch drainage networks that were constructed for this purpose over the past century altered the hydrology of the whole region. These drainage systems connected numerous prairie potholes directly to streams. Since this reduced upland storage capacity, flood-flows in streams and rivers were undoubtedly altered to some extent. Many comparisons of stream-flow patterns before and after artificial drainage have been made in the prairie pothole region, some as early as 1929. Generally, the frequency and amount of stream flooding increased with drainage of lakes and wetlands.

Agricultural engineers at Iowa State University conducted a series of

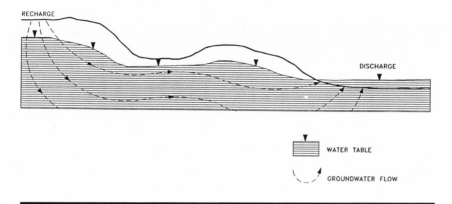

Figure 2.4. The relationship between topography, water table, and prairie pothole hydrology.

studies to estimate the effects of depressional drainage on flood-flows of a small stream in northern Iowa (Haan and Johnson, 1968a, 1968b). They found that although artificial drainage should not affect peak stream-flows from high-intensity storms, flood peaks from long duration, low-intensity storms should be greater in tiled areas. Undrained or poorly drained areas have lower peak flows, in general, but they maintain flows over a longer period. Since tile systems do not efficiently remove the water from high-intensity storm events, the water ponds in the tiled basin and drains slowly, as would an untiled wetland. In contrast, water from low-intensity storms is efficiently removed from tiled wetlands because these drainage systems are designed to carry low-magnitude flood events (those occurring every two to three years) whereas water from the same storm would remain ponded in an undrained basin.

The relationship between stream flood-flows and wetland areas may not be adequate to predict the effects of restoring wetlands on flooding. Stream channelization, which also affects flood-flows of downstream areas, occurred simultaneously with prairie pothole drainage. Straightening stream courses permits water to flow more rapidly to downstream areas because its path has been smoothed and shortened. So historical comparisons cannot distinguish between the impacts of wetland drainage and stream channelization. Moore and Larson (1979) created flood-flow models that can distinguish the effects of channelization from agricultural drainage. They found that peak stream-flows could be more highly modified by straightening stream channels than by draining depressions. Flood records for the Minnesota River support Moore and Larson's conclusions. Channelized tributaries of the river that pass through drained agricultural land have a high incidence of flooding (Moore and Larson, 1979). Flood waves from these tributaries should amplify stream-flows as the water moves downstream through the Minnesota River causing major stream flooding, but instead flows diminish as the water passes through 180 miles of unmodified stream channel. Floodwaters entering the unmodified section of the Minnesota River apparently move more slowly as they travel through the winding riverbed bordered by aquatic vegetation. The winding path increases the distance the floodwaters must travel; the aquatic vegetation gives the stream bed roughness, or obstacles, that water must flow over, around, and through.

Agricultural Runoff

Prairie soils naturally contain nitrogen and phosphorus, some of which was transported by water or sediment movement into prairie potholes, even before uplands were converted to cropland. However, nutrient-rich agricultural runoff entering wetlands has greatly increased annual inputs of nutrients since European settlement. As early as the 1940s, elevated nitrate concentrations were detected in at least one major river of the region, the Des Moines River (Keeney and DeLuca, 1993). The southern prairie pothole region is among the most productive areas in the United States for corn and soybeans, and this high productivity is the result in part of the high rates of fertilizers and pesticides applied. Agricultural applications of fertilizers and pesticides nearly doubled between 1960 and the mid-1980s, creating rural health and environmental problems (Gianessi et al., 1986; Hallberg, 1985). In recent years, fertilizer and pesticide use seems to be declining because of changing agricultural practices.

Significant amounts of fertilizers and pesticides applied to agricultural land are lost to surface water and groundwater (Hallberg, 1985). Ten to 50% of applied nitrogen and 5% of phosphorus applied are lost in agricultural drainage (Neely and Baker, 1989). Even without added inputs of fertilizer, the natural process of nitrogen mineralization, which converts organic nitrogen to nitrate, is greater in cultivated soils than in native prairie (Keeney and DeLuca, 1993). Nitrates have reached levels exceeding the drinking water standard (10 ppm) established by the Environmental Protection Agency in 5.6% of wells tested in north-central Iowa. Pesticides have been detected in 11.3% of the wells in this area (State-wide Rural Well-water Survey, 1990). Pesticides are suspected to have caused the decline of some aquatic species including reptiles and amphibians that rely on insects for food, although no experimental studies have confirmed this.

Wetlands are known to remove nutrients from water passing through them. In a study of Eagle Lake, a prairie glacial marsh in north-central Iowa with a large agricultural watershed, significant amounts of nitrate (86%) and ammonium (78%) were removed from the marsh during a year of normal rainfall (Davis et al., 1981). Although the marsh served as a sink for nitrate and ammonium, considerably less phosphate-P was removed from surface waters (20%). Nutrients in water entering wetlands may be removed because they are bound to sediments, taken up by microbes and plants, or transformed to a gas and lost to the atmosphere in the case of nitrates. However, the capacity for sediment to bind nutrients is limited; so wetlands are not always effective at removing nutrients like phosphate-P that bind to sediments (Richardson, 1985). Some of the nutrients taken up by microbes and plants also will be incorporated into wetland sediments when these organisms die. A more detailed account of the fate of nutrients in prairie wetlands is found in Neely and Baker (1989).

Plants and Animals

Less than 50 years after government surveyors passed through Iowa, Minnesota, and South Dakota to mark sections and townships for settlement in the 1850s and 1860s, the effects of market hunting and land-use changes were observed to have caused significant declines in game species, as reported by early settlers. Field naturalists began making descriptions about the status of other plants and animals after the turn of the century. Certainly, some species would have disappeared from the prairie pothole region simply because their populations were small or peripheral. As in other parts of the globe, it was the elimination of their habitat that was primarily responsible the local extinction of some species and the significant decline in populations of many other species.

Studies in northern Iowa wetlands from 1896 to 1952 have documented some major changes in the vegetation of prairie potholes. Marsh plants such as white top (*Scolochloa festucacea*) and wild rice were once widespread in marshes but have been replaced by narrow leaf cattail, hybrid cattail (*Typha glauca*), and reed canary grass (*Phalaris arundinacea*). Until the 1920s, cattails were reported only sporadically and hardly ever as dominants. Several biologists noted that only common cattail grew in prairie marshes in the early 1900s (Shimek, 1896, 1915; Wolden, 1932). By the early 1930s, narrow leaf cattails began to invade the region from the east coast (Hayden, 1939). Today, *Typha glauca,* the putative hybrid of narrow leaf and broad leaf cattail, is the dominant emergent in many prairie potholes.

Eighty-five wetland plant species are only known from a few locations within the region (Appendix A2). While some of these species had restricted ranges prior to European settlement, many are known to have declined in abundance in the past 50 to 70 years. Fens have many rare plants, but this wetland type has always been rare. Many other species that are rare today, such as some sedges and orchids, are wet prairie and sedge meadow species. These wetland types were ubiquitous but, because they were easily drained and converted to cropland, are now uncommon. Many submersed aquatics, particularly pondweeds, are also considered rare today while they were formerly common. These species have declined in part because of habitat loss, but also because of the degradation of the water quality in the remaining lakes and deeper wetlands due to increased inputs of nutrients and sediment.

Thirty species of animals are considered rare or declining in the southern prairie pothole region. The losses of large undisturbed tracts of prairie and wetland are likely responsible for the decline of species that were among the first to decline after settlement: whooping cranes (*Grus americana*), trumpeter swans (*Cygnus buccinator*), marbled godwits (*Limosa fedoa*), long-billed curlews (*Numenius americanus*), sandhill cranes (*Grus canadensis*), and common loons (*Gavia immer*) bred in undisturbed prairie marshes and associated uplands until the 1880s or 1890s. Habitat fragmentation as well as market hunting caused these species to be driven from the area by the turn of the century (Weller, 1979; Dinsmore, 1981). Animals may have specific minimum area requirements because they defend large territories, require isolation to breed, forage over extensive areas, depend on food resources that fluctuate in abundance, or depend on food resources that have area requirements. Sandhill cranes require isolated shallow wetlands with minimal human disturbance for nest sites. Sandhill cranes were once widely distributed in North America. Within the southern prairie pothole region, they were reported to have bred near Loon Lake in Jackson County, Minnesota, and across northern Iowa. The last historic Iowa nesting record was in Hancock County in 1894 (Anderson, 1907). The first sandhill crane nest since 1894 near the region was reported in eastern Iowa (Tama County) in 1992 (Poggensee, 1992).

The trumpeter swan, the largest waterfowl species in North America, is intolerant of crowding, protecting territories of 100 acres or more. Within this area, they nest and feed in the shallow water of marshes and lakes with emergent vegetation. Trumpeter swans require rather shallow water for foraging because they dig for invertebrates and plant roots. Trumpeter swans no longer breed or migrate through the southern prairie pothole region, preferring more secluded western wetlands. Trumpeter swans were reintroduced within the southern prairie pothole region in the mid-1960s to one site in Minnesota and several other sites since then. Plans to establish a wild breeding population of at least 20 pairs of trumpeter swans in northern Iowa will begin to be carried out in 1994 (Andrews and Zenner, 1992).

Whooping cranes require isolation for nest success and appear to have a very large home range. These cranes depend on extensive marsh complexes with many shallow ponds containing emergent vegetation. The whooping crane is a federal endangered species that has been extirpated from much of its former range. In 1894, a field naturalist in Iowa noted that whooping cranes were rarely observed passing over although they had been "not uncommon and bred here in earlier times" (Anderson, 1894).

Marbled godwits, willets, and long-billed curlews were common prairie species before the turn of the century. The decline of marbled godwits and willets likely resulted from market hunting and prairie conversion as well as wetland loss. Both species currently breed in the northern Great Plains, as close as northeastern South Dakota and northwestern Minnesota. These species no longer breed in Iowa. Long-billed curlews fed in shallow lakes and marshes. Birds nested on prairie near these wetland feeding areas. The town of Curlew in Pocahontas County, Iowa, serves as a reminder of its historic presence in central Iowa and the rest of the region. Long-billed curlews, once abundant in eastern South Dakota, now only breed in the western part of that state.

Common loons nested in emergent vegetation as far south as the lakes and large marshes of northern Iowa before 1900. The last known loon nest in Iowa was reported from Rice Lake in 1883 (Anderson, 1907). This species no longer breeds within the southern prairie pothole region (Hands et al., 1989c). Several other open water species have also declined. Throughout the region black terns (*Chlidonias niger*) are common nesters that seem to have declined within the past few years (Hands et al., 1989a). This species is also a migratory species of special concern to the U.S. Fish and Wildlife Service. Specific factors contributing to the decline of the species are unknown. Forster's terns, colonial nesters of very large marshes, have also declined in numbers. Horned grebes have not been documented to nest within Iowa during this century but were known to breed at Heron Lake, Minnesota. Grebes once bred throughout Minnesota but are now restricted to the northwestern portion of the state (Janssen, 1987). Current breeding sites are 200 miles north of the southern prairie pothole region. Horned grebes, also historically widespread in eastern South Dakota, have been recently rediscovered breeding in northeastern South Dakota.

Losses of emergent wetlands, both small and extensive, have affected many species. White-faced ibises (*Plegadis chihi*) feed in shallow somewhat open water. Nesting white-faced ibis reports are sporadic within the region. White-faced ibises nested on Heron Lake in Jackson County a few years between 1895 and 1939 (Janssen, 1987). More recently, ibises nested in Dickinson County, Iowa, in 1986. Likewise, common moorhens were common in Minnesota and Iowa into the 1930s, but their numbers have declined. This species requires large shallow marshes with diverse emergent vegetation. Least bitterns prefer large marshes with extensive stands of emergent vegetation and have scattered small populations throughout the region (Hands et al., 1989d). Blanding's turtle, a threatened species in Minnesota and South Dakota, prefers shallow and densely vegetated marshes.

The conversion of wet prairies and sedge meadows affected some animal species. King rails (*Rallus elegans*) have become uncommon in the southern prairie pothole region although the species was common into the 1930s. Wilson's phalarope (*Phalaropus tricolor*) breeds in sedge meadows and wet prairie next to large prairie potholes. Heron Lake in Jackson County, Minnesota, was reported to have hundreds of nesting pairs of Wilson's phalaropes before 1900 (Roberts, 1932). The species was apparently abundant in large wetlands and shallow lakes. Wilson's phalaropes suddenly disappeared around the turn of the century. Reasons for their decline are not clear. Wilson's phalaropes still breed in three counties in southern Minnesota and throughout South Dakota. The species is a rare breeder in Iowa. The two wetland-prairie raptors of the region also require a significant habitat area because they feed and defend large tracts. The northern harrier and short-eared owl, both considered endangered in Iowa, have few recent nest records, presumably because of lack of habitat (Hands et al., 1989b).

More recently, the effects of agricultural chemicals have been implicated in species declines. Specifically, insecticides are suspected to affect animals that feed on invertebrates. Smooth green snakes and pygmy shrews are two rare insectivorous species that have declined presumably because of lowered or tainted food supplies. American bitterns have suffered a loss of habitat as sedge meadows have been drained. However, breeding has declined in southern Minnesota within recent years, even in apparently suitable habitat (Coffin and Pfannmuller, 1988). A coincidental decline in frogs, an important food for bitterns, may be a factor.

Amphibian numbers within the region have precipitously declined. For example, leopard frog numbers are two to three orders of magnitude lower than at the turn of the century (Lannoo et al., 1993). Within the region, cricket frogs, once widespread, occur sporadically and mudpuppies appear to be extirpated from at least the prairies of Iowa. Several fisheries management practices have likely taken a great toll on amphibian populations. First, bullfrogs (*Rana catesbeiana*) were introduced for bait and caused native amphibians to decline in numbers. Secondly, some wetlands are used to rear game fish. Frogs and salamanders can breed in temporary and seasonal wetlands in years with at least normal precipitation. However, in drought years, frogs and salamanders must rely solely on large semi-permanent wetlands and lakes to reproduce. These small lakes and large wetlands in the region are now game fish nurseries. Rough fish are eradicated from them with rotenone in the spring, and this kills the tadpole populations of native amphibians.

Waterfowl Populations

Some species such as the mallard and blue-winged teal have been relatively successful at breeding in wetlands in agricultural landscapes, whereas other duck species have been more sensitive to land conversion. For example, northern pintails have undergone the most dramatic decline of any of North America's ducks (Ducks Unlimited, 1990). At one time, pintails were probably almost as common a breeder in the region as mallards (Roberts, 1932). Prior to 1898, the pintail was a common breeder at Heron Lake, Jackson County, in southern Minnesota, but then its numbers declined rapidly (Roberts, 1932). The northern pintail was listed as a former breeder in Iowa by 1907 (Anderson, 1907). The decline in quality of breeding grounds and the loss of ephemeral ponds are the factors most likely limiting northern pintail populations. Pintail nests are often located far from water in low-growing vegetation. So, hens on nests and broods traveling overland to water are especially susceptible to predation and other hazards in agricultural fields. Of the 12 most common prairie breeding ducks in North America, pintail numbers have been the most closely correlated with the number of flooded ponds on the prairie (Ducks Unlimited, 1990). Pintails favor ephemeral ponds and respond to a regional lack of water by continuing to search elsewhere for areas with adequate water. Since the southern prairie pothole region is wetter than the rest of the glaciated midcontinent, ephemeral ponds in the area were less affected by periodic drought. Northern pintails increased in abundance in the southern prairie pothole region during the drought of the 1930s (Weller, 1979). Since then, ephemeral ponds have been virtually eliminated from the landscape of the southern prairie pothole region by agricultural drainage. Northern pintails have lost their most reliable breeding habitat, once an important refuge from droughts to the north and west.

The gadwall (*Anas strepera*) is another waterfowl species that once was a very common breeder in the region but, as with the pintail, is now mostly a migrant. Gadwalls nest in dense, grassy vegetation on dry ground near wetlands. In 1879, gadwalls were reported to be "quite as abundant as the mallard if not more numerous" in central Minnesota (Roberts, 1932). Roberts also reported that this species is one "that suffered most severely from the settling of the country, probably as much from breaking-up of the prairie, where it commonly nested, as from the hunters." Not surprisingly, gadwalls were only reported to be an unusual nester in Iowa soon after the turn of the century (Anderson, 1907). Currently, gadwalls comprise about 1% of the breeding duck population in western Minnesota (Green and Janssen, 1975).

Canvasbacks and redheads bred in great numbers on the largest marshes and lakes, such as Heron Lake in southern Minnesota, but were essentially extirpated as breeding species from northern Iowa and southern Minnesota by the turn of the century (Roberts, 1932). Initially, overhunting probably depleted canvasback populations. The decline of habitat quality, especially the loss of extensive wild celery beds, ultimately prevented both species from successfully breeding in the region.

Unlike redheads and canvasbacks, which nest over water, lesser scaup (*Aythya affinis*) may nest on dry ground near marshes, on beds of aquatic plant stubble, or on small islands. This species only migrates through the region now but at one time was reported to have been a common breeder. Few breeding scaup have been sighted in recent years (J. J. Dinsmore, personal communication). This species now breeds chiefly in western Canada, the west coast of Hudson Bay, and southern Alaska.

Ruddy ducks are small diving ducks with nesting habits like that of redheads and canvasbacks; they nest in emergent plants over 1–3 feet of water. They will, however, use aquatic grasses and sedges for nesting. Ruddy ducks nest throughout much of North America, although they breed mainly in prairie potholes of central and western Canada and in north-central United States. Although ruddy ducks were never a preferred market hunting species, they nearly disappeared at the turn of the century from Iowa and western Minnesota. Nesting had reestablished to some extent from Lincoln County to Kittson County in Minnesota by the late 1920s (Roberts, 1932) and in northwest Iowa in the late 1930s (Low, 1941).

WATERFOWL PREDATION

Predators have historically been a source of mortality for prairie waterfowl, and the fragmentation of the prairie landscape by settlement and agricultural use has likely increased the rate of encounters between predators and waterfowl and promoted shifts in composition of the predator community (Fritzell, 1989). Predation by the red fox (*Vulpes vulpes*) is generally the major source of nest loss in dabbling ducks. Other predators including the striped skunk (*Mephitis mephitis*), raccoon (*Procyon lotor*), badger (*Taxidea taxus*), coyote (*Canis latrans*), long-tailed weasel (*Mustela frenata*), and American crow (*Corvus brachyrhnchos*) cause less predation of upland nests. Species that nest over water tend to have greater nest success because they are not as susceptible to these predators.

Since settlement, red fox populations have increased in abundance and expanded their range westward into areas formerly occupied by coyotes (Sargeant, 1982). Red foxes avoid denning in areas frequented by coyotes. Since coyotes have larger, less exclusive territories than red foxes, they do not hunt their area as thoroughly and consequently impose less mortality on nesting hens and their eggs. Red fox territories range from 1 to 3 square miles, supporting a family consisting of a mated pair and four to six young (Fritzell, 1989). Denning season coincides with nesting of prairie waterfowl. Foxes also feed on a variety of animals in upland habitats.

A series of studies at Union Slough National Wildlife Refuge in northern Iowa (Fleskes and Klaas, 1991; Gallagher, 1990) documented the effects of mammalian predators, primarily the red fox, on waterfowl production. Union Slough is a linear 912-acre wetland with minimal upland habitat, surrounded by agricultural land. Overall nest success of ground-nesting waterfowl was approximately 11%. Predators (primarily mammalian) caused 89% of these nest failures. Based on studies in North Dakota, nest success rates of 15% for mallards and 20% for other dabbling ducks are minimal to sustain breeding populations. So Union Slough actually is a sink rather than a source of waterfowl, at least in some years. When mammalian predators were continually trapped throughout the nesting season, nesting success was increased above the minimal rates needed to maintain the population, becoming a source rather than sink for regional waterfowl populations.

Habitat type or quality was not directly a factor in nest success at Union Slough. However, the extensive interface of the refuge with adjacent farmland likely causes nest densities to be high on the refuge. Predator activity is likely greater near wetlands. Expanding permanent vegetative cover around the refuge would likely disperse nests and improve nesting success, unless predator populations also increased. Similar predictions have been made for other portions of the prairie pothole region (Klett et al., 1988). Predator populations may not increase if newly established nesting habitat is included in existing predator territories (Duebbert et al., 1981). In South Dakota, nests within agricultural fields planted to permanent cover still had lower rates of destruction due to predation 10 years later (Duebbert and Lokemoen, 1976). Likewise, some dabbling ducks had higher nest success on Conservation Reserve Program lands in North Dakota and Minnesota than on Waterfowl Production Areas (Kantrud, 1993). Wide separation of nests on upland Conservation Reserve Program tracts probably reduces mammalian predation. In contrast, nests in Waterfowl Production Areas are concentrated around wetlands, prime hunting areas for both wetland and upland-oriented mammals.

Some waterfowl species rely on nest concealment as an antipredator strategy while others rely on nest dispersal. Pintails are "nest-dispersers": they tend to have nests located relatively far from breeding ponds and do not seek out dense vegetation for nest sites. Northern pintails often nest on closely grazed grasslands while the mallard, gadwall, and American wigeon prefer nest sites that are well concealed in tall, dense vegetation. Gadwalls avoid tilled uplands in favor of undisturbed perennial grasslands. Mallards use a variety of habitats other than dense grasslands for nesting, including cattails over shallow water, road right-of-ways, or cropland. The blue-winged teal, green-winged teal, and northern shoveler seem to prefer medium-height grasses for nesting but will nest in heavy cover.

Taller vegetation should provide better concealment from avian predators, but some, such as crows, cue in to nests by watching nesting hens. Vegetation more than 8 to 20 inches tall provides only slightly more protection from crows (Sullivan and Dinsmore, 1990). Duck nests farther than 0.5 mile from crow nests will be relatively safe from crow predation. Duck nests concentrated in short vegetation are more vulnerable to crow predation than those dispersed and/or in medium to tall vegetation (Sudgen and Beyersbergen, 1986, 1987).

Duckling losses also occur from predation as broods move from the nest sites to rearing areas. Red fox and other upland predators account for most mortality when broods walk overland to rearing habitat. The distance that waterfowl travel from nest sites to water varies with species. The mallard, gadwall, and pintail select nest sites up to 2 miles from water while the blue-winged teal and northern shoveler generally nest within 0.5 mile of water. Duckling losses may be greatest during overland travel when nest sites are not close to brood-rearing habitat. Raccoons, minks, and snapping turtles (*Chelydra serpentina*) will take ducklings as they travel through watercourses or after they reach rearing areas. Raccoons have increased in abundance and expanded their range since settlement (Fritzell, 1989). However, studies at Union Slough suggest brood mortality is about the same as in other prairie wetland areas although nest mortality is much higher.

REFERENCES

Anderson, R.M. 1894. The nesting of the whooping crane. Oologist 11: 263–264.

Anderson, R.M. 1907. The birds of Iowa. Proceedings of the Davenport Academy of Science 11: 125–417.

Andrews, R. and G. G. Zenner. 1992. Trumpeter swan restoration proposal. Unpublished report, September 28. Iowa Department of Natural Resources, Clear Lake. 14 pp.

Birdsall, B.P. 1915. History of Wright County, Iowa: its peoples, industries and institutions. B.F. Bowen and Company, Indianapolis, Indiana.

Bowles, J.B. 1975. *Distribution and Biogeography of Mammals of Iowa.* Special publication of the museum, Texas Tech University, Lubbock No. 9. 184 pp.

Brown, M. and J.J. Dinsmore. 1986. Implications of marsh size and isolation for marsh management. Journal of Wildlife Management 50: 392–397.

Campbell, K.L. and H.P. Johnson. 1975. Hydrological simulation of watersheds with artificial drainage. Water Resources Research 11(1): 120–126.

Christiansen, J.L. and R.M. Bailey. 1990. The *Snakes of Iowa.* Iowa Department of Natural Resources. Nongame Technical Series No. 1. 16 pp.

Christiansen, J.L. and R.M. Bailey. 1991. *The Salamanders and Frogs of Iowa.* Iowa Department of Natural Resources. Nongame Technical Series No. 3. 24 pp.

Coffin, B. and L. Pfannmuller. 1988. *Minnesota's Endangered Flora and Fauna.* University of Minnesota Press, Minneapolis.

Cowardin, L.M., D.S. Gilmer, and C.W. Shaiffer. 1985. Mallard recruitment in the agricultural environment of North Dakota. Wildlife Monograph 92: 1–37.

Cowardin, L.M., V. Carter, F.C. Golet, and E.T. LaRoe. 1979. Classification of wetlands and deepwater habitats of the United States. U.S. Fish and Wildlife Service. FWS/OBS-79/31. Washington, D.C.

Davis, C.B., J.L. Baker, A.G. van der Valk, and C.E. Beer. 1981. Prairie pothole marshes as traps for nitrogen and phosphorus in agricultural runoff. IN B. Richardson (Ed.) *Proceedings of Midwest Conference on Wetland Values and Management.* Freshwater Society, Navarre, Minnesota.

Dinsmore, J.J. 1981. Iowa's avifauna: changes in the past and prospects for the future. Proceedings of the Iowa Academy of Science 88: 28–37.

Dinsmore, J.J. 1992. Personal communication, animal ecology department, Iowa State University, Ames.

Dinsmore, J.J., T.H. Kent, D. Koenig, P.C. Peterson, and D.M. Roosa. 1984. *Iowa Birds.* Iowa State University Press, Ames. 356 pp.

Drewien, R.C. and P.F. Springer. 1969. Ecological relationships of breeding blue-winged teal to prairie potholes. Canadian Wildlife Service Report Series 6.

Ducks Unlimited. 1990. Population recovery strategy for the northern pintail.

Ducks Unlimited, Long Grove, Illinois. DU #422A. 30 pp.

Duebbert, H.F. and A.M. Frank. 1984. Value of prairie wetlands to duck broods. Wildlife Society Bulletin 12: 27–34.

Duebbert, H.F. and J.T. Lokemoen. 1976. Duck nesting in fields of undisturbed grass-legume cover. Journal of Wildlife Management 40: 39–49.

Duebbert, H.F. and J.T. Lokemoen. 1980. High duck nesting success in a predator reduced environment. Journal of Wildlife Management 44: 428–437.

Duebbert, H.F., E.T. Jacobson, K.F. Higgins, and E.B. Podoll. 1981. Establishment of seeded grasslands for wildlife habitat in the prairie pothole region. U.S. Fish and Wildlife Service Special Scientific Report 234. U.S. Department of the Interior, Washington, D.C.

Dwyer, T.J., G.L. Krapu, and D.M. Janke. 1979. Use of prairie pothole habitat by breeding mallards. Journal of Wildlife Management 43: 526–531.

Evans, C.D. and K.E. Black. 1956. Duck production studies on the prairie potholes of South Dakota. U.S. Fish and Wildlife Service Special Scientific Report 32.

Fleskes, J.P. and E.E. Klaas. 1991. Dabbling duck recruitment in relation to habitat and predators at Union Slough National Wildlife Refuge, Iowa. U.S. Fish and Wildlife Service, Fisheries and Wildlife Technical Report 32. 19 pp.

Flickinger, R.E. 1904. *The Pioneer History of Pocahonta County, Iowa, from the Time of Its Earliest Settlement to the Present Time.* Sanborn Publishers, Fonda, Iowa.

Fritzell, E. 1989. Mammals in prairie wetlands. IN A.G. van der Valk (Ed.). *Northern Prairie Wetlands.* Iowa State University Press, Ames.

Galatowitsch, S.M. 1993. Site selection, design criteria, and performance assessment for wetland restorations in the prairie pothole region. Ph.D. Dissertation, Iowa State University, Ames.

Gallagher, J.A. 1990. Predator management and effect on nesting success and recruitment of upland nesting waterfowl. M.S. Thesis, Iowa State University, Ames.

Gianessi, L.P., H.M. Peskin, P. Crosson, and C. Puffer. 1986. Nonpoint source pollution: are cropland controls the answer? Report prepared for the U.S. Environmental Protection Agency, USDA Soil Conservation Service, and U.S. Geological Society under EPA Cooperative Agreement CR 811 858 01. Resource for the Future. Washington, D.C.

Great Plains Flora Association. 1986. *Flora of the Great Plains.* University Press of Kansas, Lawrence. 1392 pp.

Green, J.C. and R.B. Janssen. 1975. *Minnesota Birds.* University of Minnesota Press, Minneapolis.

Haan, C.T. and H.P. Johnson. 1968a. Hydraulic model of runoff from depressional areas, part I. General considerations. Transactions of the American Society of Agricultural Engineers 11(3): 364–367.

Haan, C.T. and H.P. Johnson. 1968b. Hydraulic model of runoff in depressional areas, part II. Development of the model. Transactions of the American Society of Agricultural Engineers 11(3): 368–373.

Hallberg, G.R. 1985. Groundwater quality and agricultural chemicals: a perspective from Iowa. Proceedings of the North Central Weed Control Conference, Iowa Fertilizer and Chemical Association. Des Moines, Iowa. 36 pp.

Hallberg, G.R. 1986. Nitrates in groundwater in Iowa. Proceedings of the Nitrogen and Groundwater Conference, Iowa Fertilizer and Chemical Association. Des Moines, Iowa. 36 pp.

Hands, H.M., R.D. Drobney, and M.R. Ryan. 1989a. Status of the black tern in the northcentral United States. Report prepared for the U.S. Fish and Wildlife Service, Twin Cities, Minnesota. 15 pp.

Hands, H.M., R.D. Drobney, and M.R. Ryan. 1989b. Status of the northern harrier in the northcentral United States. Report prepared for the U.S. Fish and Wildlife Service, Twin Cities, Minnesota. 18 pp.

Hands, H.M., R.D. Drobney, and M.R. Ryan. 1989c. Status of the common loon in the northcentral United States. Report prepared for the U.S. Fish and Wildlife Service, Twin Cities, Minnesota. 26 pp.

Hands, H.M., R.D. Drobney, and M.R. Ryan. 1989d. Status of the least bittern in the northcentral United States. Report prepared for the U.S. Fish and Wildlife Service, Twin Cities, Minnesota. 13 pp.

Hayden, A. 1939. Notes on *Typha angustifolia* in Iowa. Iowa State Journal of Science 13(4): 341–352.

Hazard, E.B. 1982. *The Mammals of Minnesota*. University of Minnesota Press, Minneapolis. 280 pp.

Helmers, D.L., M.R. Ryan, and L.H. Frederickson. 1990. Management of shorebird habitat for invertebrate availability in the Great Plains. IN *Proceedings of the Non-Game Migratory Bird Workshop*. U.S. Fish and Wildlife Service Office of Information Transfer, Fort Collins, Colorado.

Hewes, L. 1951. The northern wet prairie of the United States: nature, sources of information, and extent. Annals of the Association of American Geographers 41: 307–323.

Iowa State Census Reports. 1925. State of Iowa Historical Society files. Des Moines.

Janssen, R.B. 1987. *Birds in Minnesota*. University of Minnesota Press and James Ford Bell Museum of Natural History, Minneapolis. 352 pp.

Jones, J.K., Jr., D.M. Armstrong, and J.R. Choate. 1985. *Guide to Mammals of the Plains States*. University of Nebraska Press, Lincoln. 371 pp.

Kaminski, R.M. and H.H. Prince. 1981. Dabbling duck and aquatic macroinvertebrate response to manipulated wetland habitat. Journal of Wildlife Management 45: 1–15.

Kantrud, H.A. 1993. Duck nest success on Conservation Reserve Program land in the prairie pothole region. Journal of Soil and Water Conservation 48(3): 238–242.

Kantrud, H.A. and R.E. Stewart. 1977. Use of natural basin wetlands by breeding waterfowl in North Dakota. Journal of Wildlife Management 41: 243–253.

Kantrud, H.A., J.B. Millar, and A.G. van der Valk. 1989. Vegetation of wetlands of the prairie pothole region. IN A.G. van der Valk (Ed.), *Northern Prairie Wetlands*, Iowa State University Press, Ames.

Keeney, D.R. and T.H. DeLuca. 1993. Des Moines River nitrate in relation to watershed agricultural practices: 1945 versus 1980s. Journal of Environmental Quality 22:267–272.

Kemmis, T.J. 1991. Glacial landforms, sedimentology, and depositional environments of the Des Moines Lobe, northern Iowa. Ph.D. Dissertation, University of Iowa, Iowa City. 393 pp.

Klett, A.T., T.L. Shaffer, and D.H. Johnson. 1988. Duck nest success in the prairie pothole region of the United States. Journal of Wildlife Management 52: 431–440.

LaGrange, T.G. and J.J. Dinsmore. 1989. Habitat use by mallards during spring migration through central Iowa. Journal of Wildlife Management 53: 1076–1081.

Lannoo, M.J., K. Lang, T. Waltz, and G.S. Phillips. 1993. Profile of an abraded amphibian assemblage: Dickinson County, Iowa, 70 years after Frank Blanchard's survey. Unpublished manuscript.

Lokemoen, J.T. 1973. Waterfowl production on stock-watering ponds in the northern plains. Journal of Range Management 26: 179–184.

Low, J.B. 1941. Nesting of the ruddy duck in Iowa. Auk 58: 506–517.

Moore, I.D. and C.L. Larson. 1979. Effects of drainage projects on the surface runoff from small depressional watersheds in the north central region. Water Resources Research Center Technical Bulletin 99. Minneapolis, Minnesota. 218 pp.

Murkin, H.R., R.M. Kaminski, and R.D. Titman. 1982. Responses by dabbling ducks and aquatic invertebrates to an experimentally manipulated cattail marsh. Canadian Journal of Zoology 60: 2324–2332.

Neely, R.K. and J.L. Baker. 1989. Nitrogen and phosphorus dynamics and the fate of agricultural runoff. IN A.G. van der Valk (Ed.), *Northern Prairie Wetlands,* Iowa State University Press, Ames.

Nelson, J.W. and J.A. Kadlec. 1984. A conceptual approach to relating habitat structure and macroinvertebrate production in freshwater wetlands. Transactions of North American Wildlife and Natural Resources Conference 49: 262–270.

Poggensee, D. 1992. Nesting sandhill crane at Otter Creek Marsh, Tama County, Iowa. Iowa Bird Life 62: 112–113.

Reid, F.A., W.D. Rundle, M.W. Sayre. 1983. Shorebird migration chronology at two Mississippi River Valley wetlands of Missouri. Transactions of the Missouri Academy of Science 17:103–115.

Richardson, C.J. 1985. Mechanisms controlling phosphorus retention capacity in freshwater wetlands. Science 228: 1424–1427.

Roberts, T.S. 1932. *Birds of Minnesota.* University of Minnesota, Minneapolis. Vols. 1 and 2.

Ruwalt, J.J., Jr., L.D. Flake, and J.M. Gates. 1979. Waterfowl paie use of natural and man-made wetlands in South Dakota. Journal of Wildlife Management 43:375–383.

Sargeant, A.B. 1972. Red fox spatial characteristics in relation to waterfowl predation. Journal of Wildlife Management 36: 225–236.

Sargeant, A.B. 1978. Red fox prey demands and implications to prairie duck production. Journal of Wildlife Management 42: 520–527.

Sargeant, A.B. 1982. A case history of a dynamic resource—the red fox. IN G.C. Sanderson (Ed.) *Midwest Furbearer Management.* Proceedings of the 1981 symposium, Midwest Fish and Wildlife Conference, Wichita, Kansas, North Central Section, Central Mountains and Plains Section, and Kansas Chapter, the Wildlife Society.

Shimek, B. 1896. Notes on aquatic plants from northern Iowa. Proceedings of the Iowa Academy of Science 4: 77–81.

Shimek, B. 1915. The plant geography of the Lake Okoboji Region. Laboratory of Natural History — University of Iowa — Bulletin 7(2): 1–90.

Sousa, P.J. 1985. Habitat suitability index models: gadwall (breeding). U.S. Fish and Wildlife Service Biological Report 82 10-100. 35 pp.

South Dakota Ornithologists' Union. 1991. *The Birds of South Dakota.* Second Edition. Northern State University Press, Aberdeen, South Dakota. 411 pp.

State-wide Rural Well-water Survey. 1990. State of Iowa, Des Moines.

Stewart, R.E. and H.A. Kantrud. 1971. Classification of natural ponds and lakes in the glaciated prairie region. U.S. Fish and Wildlife Service Resource Publication 92. 57 pp.

Stewart, R.E. and H.A. Kantrud. 1973. Ecological distribution of breeding waterfowl populations in North Dakota. Journal of Wildlife Management 37: 39–50.

Sudgen, L.G. and G.W. Beyersbergen. 1986. Effect of density and concealment on American crow predation of simulated duck nests. Journal of Wildlife Management 50: 9–14.

Sudgen, L.G. and G.W. Beyersbergen. 1987. Effect of nesting cover density on American crow predation of simulated duck nests. Journal of Wildlife Management 51: 481–485.

Sullivan, B.D. and J.J. Dinsmore. 1990. Factors affecting egg predation by American crows. Journal of Wildlife Management 54: 433–437.

Swanson, G.A. and H.F. Duebbert. 1989. Wetland habitat of waterfowl in the prairie pothole region. IN A.G. van der Valk (Ed.). *Northern Prairie Wetlands.* Iowa State University Press, Ames.

van der Valk, A.G. and C.B. Davis. 1978. The role of seed banks in the vegetation dynamics of prairie glacial marshes. Ecology 59: 322–335.

van der Valk, A.G. and C.B. Davis. 1979. A reconstruction of the recent vegetational history of a prairie marsh, Eagle Lake, Iowa, from its seed bank. Aquatic Botany 6: 29–51.

van der Valk, A.G., C.B. Davis, J.L. Baker, and C.E. Beer. 1979. Natural freshwater wetlands as nitrogen and phosphorus traps for land runoff. IN P.E. Greeson, J.R. Clarke, and J.E. Clark (Eds.). *Wetland Functions and Values: The State of Our Understanding.* American Water Resources Association, Minneapolis, Minnesota.

Voigts, D.K. 1976. Aquatic invertebrate abundance in relation to changing marsh vegetation. American Midland Naturalist 95: 313–322.

Weller, M.W. 1979. Birds of some Iowa wetlands in relation to concepts of faunal preservation. Proceedings of the Iowa Academy of Science 86: 81–88.

Weller, M.W. and C.E. Spatcher. 1965. The role of habitat in the distribution and abundance of marsh birds. Iowa State University Agriculture and Home Economics Experiment Station Special Report No. 43.

Weller, M.W. and L.H. Fredrickson. 1974. Avian ecology of a managed glacial marsh. Living Bird 12: 269–291.

Winter, T.C. 1989. Hydrologic studies of wetlands in the northern prairie. IN A.G. van der Valk (Ed.). *Northern Prairie Wetlands.* Iowa State University Press, Ames.

Wolden, B.O. 1932. The plants of Emmet County, Iowa. Proceedings of the Iowa Academy of Science 39: 89–132.

SITE SELECTION GUIDELINES

Since wetlands historically covered over a third of the landscape of the southern prairie pothole region, drained basins occur in nearly every square mile. Except in cities, towns, and along roads, few prairie potholes have been filled—most were drained with ditches or tiles and then cultivated or used as pasture. This means that the vast majority of drained potholes potentially can be restored. In fact, the number of drained basins made available to local, state, federal, and private agencies for restoration has exceeded the resources available to renovate them. Because some drained basins are clearly better candidates for restorations than are others, guidelines for prioritizing sites are needed. For example, restoring a wetland that will not adversely affect roadbeds or agricultural use of adjacent land is preferable to restoring one that will. Guidelines for selecting basins to be restored, however, should go beyond the obvious prescriptions for avoiding sites where there might be problems with adjacent land use. Wetland restorations are being done for a variety of reasons, including simply taking land out of agricultural production, but three reasons for restoring wetlands are most commonly cited: (1) creating wildlife habitat, (2) improving water quality, and (3) reducing flooding problems. Site selection guidelines thus should include criteria for recognizing sites that have the best potential to become good wildlife habitat, to have an beneficial impact on water quality, and to provide maximum storage for flood water. Selection guidelines must also consider physical and hydrological factors to ensure that a successful restoration is possible. These include the size and maximum depth of the basin; soil types present in the basin; size of the watershed; expected inputs of nutrients, pesticides, and sediments; proximity to other natural and restored wetlands; and potential for natural revegetation. Consequently, projects should be prioritized on the basis of their potential for functioning as wildlife habitat, water purifiers, and flood storage reservoirs and their chances for being implemented successfully. We first discuss factors that should be considered when selecting wetlands to be restored for wildlife habitat, for improving water quality, or for flood storage. We then outline how to locate drained wetland basins to restore, how to conduct an assessment of a restoration site, and what permits may be needed to do a restoration project.

FUNCTIONAL GUIDELINES

When choosing sites for restoration, two different types of criteria can be used to set priorities: (1) historical criteria and (2) functional criteria. Historical criteria are based on the types of presettlement wetlands found in an area. The highest priority for restoration would be basins that formerly contained wetlands that are now the most underrepresented in the area. Functional criteria are used to select sites that will best provide a certain function, e.g., increase dabbling duck breeding habitat or nitrate removal. These two sets of criteria are not mutually exclusive, and they can be and should be combined. For example, replacing wetlands using historical criteria should result in a diversity of wetland types that will also provide a wide array of functions. Likewise, if waterfowl habitat is the primary reason for a restoration, the restoration of a complex of different types of wetlands may be required rather than just one type of wetland.

In order to prioritize restorations using historical criteria, the kinds of wetlands that are lacking in an area must first be determined. This can be done by estimating from soil maps what classes of wetlands (temporary, seasonal, semi-permanent, and permanent) were present and their abundance at the time of European settlement (see section on locating restoration sites in this chapter) and comparing this to data from contemporary wetland maps available locally from the Soil Conservation Service, U.S. Fish and Wildlife Service, and state wildlife biologists. (These agencies often maintain maps of existing wetlands that can be viewed in their offices, e.g., National Wetland Inventory maps.) If semi-permanent or permanent wetlands are absent and they were formerly present, then they should be made the highest priority for restoration. Temporary wetlands are commonly the most underrepresented wetland type because they were easily drained. Seasonal wetlands are the next priority for restoration if there is less extant acreage of them than for semi-permanent wetlands. Temporary wetlands should be the highest restoration priority if they occupy less than four times the area of either natural seasonal or semi-permanent basins. Ideally, restoration projects using historical criteria will select large tracts of land to restore and will restore all the wetlands in the tract as well as the surrounding uplands back to prairie. This is also the best approach for projects whose major goal is restoring wildlife habitat. If wetland basins are to be restored singly, then functional criteria for selecting sites should be considered.

Functional criteria for prioritizing restoration sites are more complicated because they require a detailed assessment of the site to be restored and an evaluation of how its characteristics compare to those of basins whose physical features and watershed have the desired function. Factors to be considered in assessing sites for wildlife, water quality, and flood storage are outlined next. It should be noted that all wetlands have wildlife habitat, water quality, and flood storage functions. They differ in degree from wetland to wetland. A wetland restored to provide habitat for breeding dabbling ducks may also remove some nitrates from water passing through it, but not as effectively as a wetland restored specifically to improve water quality. A wetland restored primarily for water quality purposes will also be used by wildlife, but mostly by generalists.

Wildlife Habitat

The original configuration of prairie pothole landscape had different kinds of wetland habitats within a locale—not within every basin. Trying to replicate a variety of habitats in a single basin cannot substitute for having different kinds of wetlands since many species will not use small habitat patches (Brown and Dinsmore, 1986). When choosing between several possible restoration sites or design configurations, wildlife will be best served by restoring sites in which a mix of temporary, seasonal, and semi-permanent wetlands can be established.

Minimal requirements for vertebrates are vegetative cover, adequate nesting sites, adequate food, and sometimes adequate isolation from other individuals of the same species. Nine distinctive groups of animals inhabit prairie potholes—each with unique minimal requirements. Birds with very large area requirements, generally complexes of wetlands and associated grasslands, such as trumpeter swans and marbled godwits are in Group AS (area sensitive). Birds such as gulls, terns, and diving ducks requiring large open bodies of water or associated lakeshore habitat are in Group OW (open water). Group MG includes birds that are marsh generalists, requiring some emergent vegetation—such as coots and yellow-headed blackbirds. Bitterns, rails, yellowthroats, sedge wrens, and other birds requiring development of high-quality sedge meadows and shallow emergents comprise Group SB (secretive birds). Group DG are dabbling ducks and geese. Shorebirds, mostly migratory, require mudflats for foraging (Group SH). Group RA includes reptiles and amphibians. Small mammals often associated with sedge meadows are Group SM, and fur bearers (Group FB) require deep marshes that do not freeze to the bottom in winter. A list of species in each group is given in Table 3.1.

The minimal habitat requirements for each group of animals in Table 3.1 are given in Table 3.2 and on the data sheet in Appendix D6. These minimal requirements fall into three categories: (1) basin characteristics that are more or less permanent features of the basin, such as size, duration of ponding (ephemeral to semi-permanent), and basin morphometry; (2) vegetation extent and composition; and (3) watershed land use, i.e., how much of the watershed is cropland or permanent cover. This information (Table 3.2) can be used as guidelines for selecting sites that will provide suitable habitat for a given group of animals.

Water Quality

The amount of row-crop agriculture in a watershed (Figure 3.1) is highly correlated with the concentration of nitrogen and phosphorous in its surface waters (Omernik, 1977). Restoring wetlands in agricultural watersheds to remove nutrients from agricultural runoff has great potential for reducing nonpoint source pollution problems associated with agriculture (van der Valk and Jolly, 1992). Nutrients in agricultural runoff that enters wetlands can be taken up by microbes and plants, bound to sediments, or, in the case of nitrates, transformed to a gas and lost to the atmosphere (denitrification). Nutrient removal efficiency depends on the time it takes for water to pass through the basin (residence time), how much vegetation and litter are present, water and sediment temperatures, etc. Most studies in prairie marshes and other wetlands have dealt with the fate of nitrogen, particularly nitrates, and phosphorus in wetlands, and almost nothing is known about the fate or impact of pesticides in these wetlands (Neely and Baker, 1989).

Nutrients and pesticides are transported through the landscape on sediment, in surface runoff, or in subsurface runoff. Sediment, surface runoff, and subsurface runoff carry different amounts and types of nutrients and pesticides because of soil adsorption and water solubility char-

TABLE 3.1. Groups of animals that are found in prairie potholes

GROUP	DEFINITION	SPECIES
AS	Birds with large area requirements - generally complexes of wetlands and associated grasslands.	Trumpeter swan, willet, whooping crane, sandhill crane, long-billed curlew, marbled godwit, northern harrier, short-eared owl.
OW	Birds that require large, semipermanent wetlands or lakes. Many of the birds are colonial waterbirds or fish-eating species.	Horned grebe, eared grebe, western grebe, pied-billed grebe, red-necked grebe, American white pelican, great egret, great blue heron, green-backed heron, black-crowned night heron, least bittern, white-faced ibis, redhead, canvasback, ruddy duck, ring-necked duck, common loon, Franklin's gull, Forster's tern, black tern.
MG	Marsh generalists: birds that can use smaller wetlands and require some robust emergent vegetation.	American coot, common moorhen, yellow-headed blackbird, red-winged blackbird.
SB	Secretive birds of shallow marshes including birds that require sedge meadows and wet prairie.	American bittern, least bittern, king rail, Virginia rail, sora, Wilson's phalarope, sedge wren, marsh wren, LeConte's sparrow, savannah sparrow, swamp sparrow, Henslow's sparrow, common yellowthroat.
DG	Dabbling ducks and geese - often require several kinds of marshes to complete life stages.	Canada goose, mallard, gadwall, northern pintail, green-winged teal, blue-winged teal, northern shoveler.
SH	Birds that require extensive bare soil - mud, sand, or gravel, for nesting and foraging - shorebirds.	Killdeer, common snipe, spotted sandpiper, American avocet, ca 25-30 species of migrating shorebirds.
RA	Reptiles and amphibians.	Tiger salamander, northern leopard frog, American toad, Canadian toad, chorus frog, cricket frog, western painted turtle, snapping turtle, Blanding's turtle, smooth green snake, eastern garter snake, western plains garter snake, mudpuppy, bullfrog.
SM	Small mammals - mammals inhabiting wet prairies and sedge meadows.	Masked shrew, pygmy shrew, short-tailed shrew, Franklin's ground squirrel, meadow vole, meadow jumping mouse, white-footed mouse, deer mouse.
FB	Fur bearers - mammals requiring semipermanent marshes.	Muskrat, mink, beaver, raccoon, ermine, long-tailed weasel, least weasel.

TABLE 3.2. Minimal habitat requirements for groups of animals in Table 3.1

GROUP	CHARACTERISTICS
AS: Area Sensitive Birds	At least 640 acres of adjacent land in permanent cover.
	At least 100 acres of wetland present, including a Class IV basin.
OW: Open Water Birds	Class IV or V wetlands.
	Wetland size greater than 40 acres.
MG: Marsh Generalists (birds)	Class III or IV wetlands.
	At least patches of shallow emergent and deep emergent vegetation.
SB: Secretive Birds	Continuous vegetative cover of sedge meadow and shallow emergent zones.
	Proportion of area less than 18" deep greater than 50%.
	A buffer of permanent cover surrounding basin.
DG: Dabbling Ducks and Geese	Class II, III, IV wetlands.
	Proportion of area less than 18" deep greater than 50%
	A buffer of permanent cover surrounding basin.
	Wetlands of all Classes (I-V) present in vicinity.
SH: Shorebirds	Shallow or exposed unvegetated areas.
RA: Reptiles and Amphibians	Class II, III, IV wetlands.
	A buffer of permanent cover surrounding basin.
SM: Small Mammals	Continuous vegetative cover of wet prairie and sedge meadow zones.
	A buffer of permanent cover surrounding basin.
FB: Fur bearers	Class IV wetlands.
	Well-developed vegetative cover of shallow and deep emergent zones.

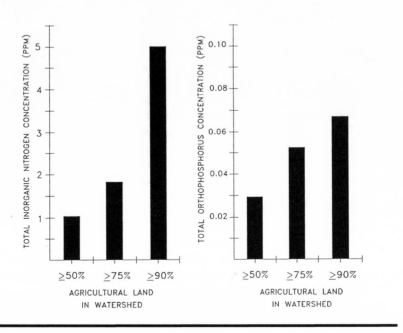

Figure 3.1. The effects of agricultural use on total inorganic nitrogen and total phosphorous concentrations in streams of the northeastern United States. (Adapted from Omernik, 1977)

acteristics of different compounds. For example, most phosphorus is transported with sediment while most ammonium-nitrogen is transported in surface runoff, and nitrate-nitrogen is transported in subsurface drainage (Neely and Baker, 1989). One of the most ubiquitous water quality problems in agricultural landscapes is high concentrations of nitrates in agricultural runoff. Wetlands with tile inputs will have significant inputs of nitrates as long as the area drained has fields that are fertilized and cultivated. Wetlands, however, can remove nitrates in incoming water very effectively as demonstrated in experimental studies in which nitrate was added to small experimental marshes at concentrations (10 ppm) similar to those in tile lines from crop fields. After five days, 80% of the nitrogen was lost through denitrification while 20% was taken up by plants or stored in litter or sediment (Isenhart, 1992). In fact, nitrate concentrations in prairie wetlands are usually below detectable limits except near tile inlets or after major storm events.

Although restored wetlands that intercept agricultural runoff from tile-drained row crops could be very important for reducing nitrate levels in prairie rivers, lakes, and reservoirs, few restored wetlands in the southern prairie pothole region actually receive much agricultural runoff. Less than one-fifth of restorations have tile inputs, and those are often at the headwaters of tile systems (Galatowitsch, 1993). The most important siting criteria for wetlands being constructed to improve water quality is selecting basins that will have subsurface agricultural runoff as their major source of water.

There is an upper limit to the nutrient- and contaminant-processing capacity of a wetland. Excessive contaminant loads may seriously degrade restored wetlands, particularly as wildlife habitat. For example, nutrient inputs can cause algal blooms, and algal blooms can cause a decline in aquatic plants, such as pondweeds, which are important wildlife food plants (Crumpton, 1989). Pesticides can reduce invertebrate populations. A restoration designed primarily to treat agricultural runoff needs to be large enough to avoid being overloaded by incoming sediments, nutrients, and contaminants. Avoid restoring wetlands in basins that will receive drainage from farmsteads, feedlots, corrals, sewage lines, and septic fields. All are concentrated sources of nutrients and/or contaminants. If these kinds of areas are within the watershed of the restoration, divert their flow from restored basins, if possible.

It has been suggested that restored marshes could contribute to groundwater contamination because nutrients and other contaminants in their water may move into the groundwater. In most cases, this is highly unlikely because nitrates are rapidly lost from surface water in wetlands. Groundwater contamination would occur only if the restored wetland had very permeable soils. Most prairie pothole soils have very fine-textured soils that have a low hydraulic conductivity (i.e., low permeability). Prairie potholes that have peat soils (with higher permeability) are usually sites of groundwater discharge, not recharge, and so should not be potential groundwater contamination sites.

The only sites that could become conduits for nitrate movement to groundwater are those that are groundwater recharge sites, i.e., sites with soils classified as "Albolls" (Figure 3.2). These soils have a subsurface layer that has been thoroughly leached of clay, iron oxides, and

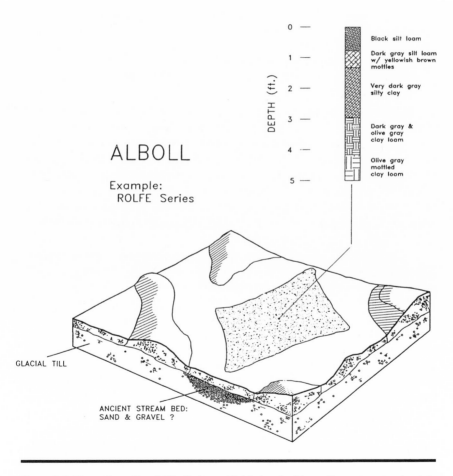

Figure 3.2. The soil profile of Albolls has a highly leached layer (here at approximately 1-foot depth), which indicates net downward movement of water.

organic matter and another layer that has a buildup of clay. These two features suggest that larger quantities of water have passed through these soils than through surrounding soils (Thompson, 1992). Channels made by plant roots and tunneling animals provide routes for downward flow of water (Thompson, 1987). Rapid downward movement of water would likely occur where there was a porous material below the soil, for instance, an ancient streambed of sand and gravel deposits. Although no studies have been done that describe if the strong horizon development in all Albolls in the prairie pothole region is the result of groundwater recharge, these sites should be avoided for restoration projects if high nitrate and pesticide loadings are possible on the sites, for example, basins with large-diameter tile lines that drain watersheds mostly in row crops.

Potential Flood-Flow Alteration

Large numbers of restored wetlands combined with stream dechannelization would be necessary to significantly alter flood-flows of large streams and rivers in this region. However, there may be beneficial effects on small streams and drainage systems that are worth considering. Since most of the small watersheds within the region discharge to channelized streams, the increased flows from agricultural drainage reach the channels. Reducing storm runoff in small streams could lessen bank erosion and downstream transport of sediments. The value of restored wetlands is primarily to small streams and drainage systems in the immediate vicinity of restoration projects.

Moore and Larson (1980) created a model that compared the alteration of flood events in waterways from stream channelization and basin drainage. They predicted the effects of flooding under four hypothetical conditions: (1) natural, with no channelized streams and no drainage of depressions; (2) intermediate, with channelized streams but no wetland drainage; (3) present (1956–1962), with channelized streams, 25% of the wetlands not drained, 50% of the wetlands drained by ditches, and 25% of the wetlands drained by tile; and (4) future, with channelized streams, 25% of the wetlands drained by ditches and 75% drained by tile. They found that changes from the "present" condition to "future" condition would increase storm runoff volumes 50% to 80%. Jacques and Lorenz (1987) used a different approach, making predictions based on the storage in the wetland basins not drained and the total stream drainage area. The Jacques and Lorenz model is simpler, because it ignores kinds of drainage and stream channelization. This model can be easily solved with available information and so was used to establish preliminary restoration guidelines (Figure 3.3). More study, especially testing model predictions, will be necessary before we can be confident of our understanding of how increasing wetland restoration acreage will affect flood-flow in a specific stream watershed. Flood-flow reductions are going to be significant, however, only in areas where multiple restorations or large restorations have been completed. The restoration of wetland complexes will have more impact than restoring individual basins within an area. In general, if the wetlands in a basin do not have the capacity to store 0.5 inch of runoff from the entire watershed, peak flows will not be reduced (Bartels, 1993).

Realigning streams into their natural channels and reestablishing fringe beds of vegetation might have a greater effect on flood-flows than restorations of depressional wetlands (Moore and Larson, 1980). This has seldom been done in the region and is beyond the scope of this book. Occasionally, restoration projects within the region have been proposed that would impound the main stem of small streams. These projects

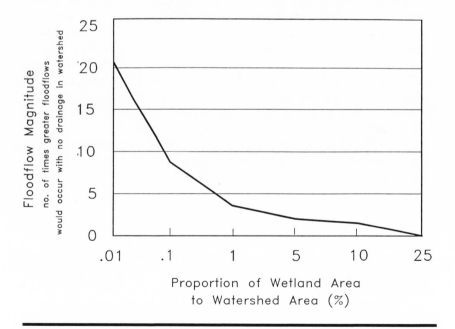

Figure 3.3. The effects of wetland drainage on flood magnitude based on predictions from a model developed by Jacques and Lorenz (1987). These results pertain to drainage areas between 1 square mile and 100 square miles for 2-year to 10-year flood events.

could actually result in environmental degradation because reduced stream-flows can have adverse effects on the downstream biota. Stream restorations that restore natural channels should be considered whenever wetland complexes are being restored.

Most flood-flow studies use specific points along water courses to evaluate storm runoff. Likewise, the most logical way to assess potential effects of wetland restoration is to select stream and drainage system "evaluation points." The appropriate stream evaluation point is where the ditch or culvert that serves as an outlet for the wetland joins a stream. If this stream is a major waterway such as the Des Moines or Minnesota Rivers, it is extremely unlikely that small marshes will have a measurable impact on flood-flows. For other cases, record the size of restored wetland basin (RW) and the size of the stream watershed (SW) above the evaluation point. Calculate the proportion of the restoration area to stream watershed size: RW/SW. If RW/SW is greater then 0.005 then flood-flow reduction is possible.

Local drainage systems also may benefit from restoration of wetlands. Removing part of the area serviced by large tile mains and county ditches may improve drainage of the remaining downstream system because many are significantly overcommitted. The effects on a local drainage system can be evaluated in the same manner. Obtain a map of the public drainage system for your area. Follow the tile or ditch line from the project area to the public waterway and then downstream until it exits the drainage district into a stream or ditch system. This is the evaluation point. If an organized drainage district services this area, use its boundary as the evaluation area. If not, calculate the drainage area based on public ditch and tile maps for the county. Record the total area of wetland restorations (RW) and the area of the drainage district boundary (DW). Calculate RW/DW and determine if the ratio exceeds 0.005.

FLOOD-FLOW ALTERATION

One of the largest wetlands restored in the region is in Redwood County, Minnesota (Figure 3.4). This basin was once called Horseshoe Lake and straddled portions of three sections. Approximately one-third, or 80 acres, of the lake was restored under the State of Minnesota's Reinvest in Minnesota Program. The watershed for the restored wetland is a portion of the watershed of Clear Creek. Before the restoration, water drained from the cropland into tiles that emptied into ditches, and finally to Clear Creek, approximately 5 miles to the northeast. Clear Creek is a tributary of the Redwood River. The Redwood River enters the Minnesota River approximately 14 aerial miles from the restored wetland.

The stream most likely to have reduced flood potential because the wetland has been restored is Clear Creek. It has a much smaller watershed than the Redwood River and received drainage directly from ditch lines that used to drain the restored wetland. The small tributary

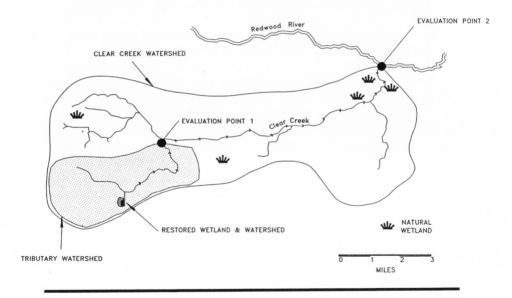

Figure 3.4. Map showing evaluation points used for calculating the effect of restoring a wetland on the flood-flow of Clear Creek in Redwood County, Minnesota. Stippled area is a watershed of a tributary of Clear Creek.

of Clear Creek (actually a drainage ditch) drains 9,600 acres (Evaluation Point 1). The drainage area was determined from the county highway map, which shows the public drainage system. There are no other wetlands in this drainage area. So the ratio of restored wetland area (RW) to drainage ditch watershed area is 0.008 (0.8%). Since this value is between 0.001 and 0.05, flood-flows in this tributary of Clear Creek could possibly be reduced slightly. Flood-flows' magnitudes will still be about three and one-half to four times greater than would be expected from a comparable area with little wetland drainage, or about 25% of the area as storage (see Figure 3.3).

This restoration project will not affect flooding along Clear Creek or the Redwood River. The Clear Creek watershed is approximately 49,280 acres (Evaluation Point 2). The storage area to drainage size (RW/SW) for this project is 0.0016. However, four other wetlands, totaling 490 acres, are within the Clear Creek watershed. So the combined effects of all wetlands in the drainage areas would be 570/49,280 or 0.011 (1.1%). Flood-flows in Clear Creek are still predicted to be approximately three times that of a largely undrained watershed.

LOCATING RESTORATION SITES

In areas that have been relatively recently drained, landowners and neighbors may recall the locations of historic wetlands. For most properties, this information is no longer available. The locations of former wetlands can still be found using county soil survey maps. In fact, soil survey maps are often the only way to locate and obtain essential information, such as size, shape, boundaries, and type of wetland formerly present on a tract of land. County soil surveys are generally available from the local Soil Conservation Service (SCS) office. Often, if a recent soil survey has not been published, a set of field maps with the best available information is maintained in the local SCS office.

Historic wetland basins in a tract of interest can be located by identifying wetland soil map units on the appropriate county soil survey map. Find map units that indicate the soil was formed in a wetland setting—Aquolls, Histosols, and Fluvaquents. Soils that formed in prairie potholes have either organic soils (peats) or mineral soils with gray subsoil. Upland soils were well-drained soils that often have a yellowish brown rather than gray subsoil. Areas with upland soils are usually not good sites for wetland restoration projects because such sites will usually not have adequate watersheds and will probably not hold standing water.

A list of wetland soil types (soil series) can be found in Appendix C. The soil series listed in this appendix include all of the major soil types of depressional wetlands in the region. (Floodplain and floodplain-associated soils — for instance on river terraces — are found throughout the region but are not within the scope of the manual.) If a putative wetland basin is mapped as a soil type that is not listed, it may be because the area was not a prairie pothole or it contains a minor wetland soil type. For these rare cases, a soil scientist will need to be consulted to determine whether a wetland formerly was present or not. In all cases, the identification of soils indicated on the soils map should be verified in the field. Take a core of soil at least 3 feet deep, and compare it to the characteristics described for the series in the county survey. A simple key to wetland soils of the region is provided in Appendix C. This key should be particularly useful in areas that do not have a recently published soil survey. SCS staff can assist in verifying soil identification if needed.

RESTORATION FEASIBILITY

The suitability of a basin for restoration for a particular function will depend primarily on (1) how much of the basin can be restored without conflicts on adjacent lands, (2) the potential hydrology of the restored wetlands, and (3) the potential for natural revegetation. Some of the information needed to make these assessments about a basin occasionally can be obtained from landowners, neighbors, or agency files. But much of this information must be acquired from a field survey or extrapolated from field data.

Extent of Basin

Making a base map is an essential first step for evaluating a basin. This base map should indicate the location, extent, and topography of the basin; land ownership in and around the basin; and drainage features of the basin, particularly the sizes of input tiles and ditches. Possible locations of dikes and water control structures should also be noted on it. The base map should include the entire basin before drainage and its watershed, even if it is not entirely restorable. The extent of the wetland basin before agricultural use can be ascertained from county surveys by drawing a line around all contiguous wetland soil units. The original extent of the basin may not be readily evident in the field because of differences in land use within the basin or because of intervening roads

or railroad grades. The portion of the basin that is restorable is determined after confirming land ownership boundaries, roads, power lines,
and other obstacles in the field.

Neighboring landowners can potentially be affected by wetland restoration projects either because their land will be flooded or because
drainage of their land will be impeded. Flooding adjacent land is a near
certainty if the basin crosses property lines. Some drainage systems (tile
or ditch) that are interrupted to restore a wetland could hinder drainage
of upstream lands even if normal pool elevation does not cross the property boundary. Special consideration should be given to sites where a
wetland project is close to a property line: a basin survey should be
conducted, water retention structures (e.g., dikes) should be low enough
to contain water on the site, and neighbors should be informed of the
project. On a few projects, to restore a wetland that could occasionally
flood adjacent property, arrangements have been made to compensate
the owner for losses. These agreements are sometimes called flowage
easements.

The topographic survey needs to be of sufficient detail to determine
the maximum pool elevation permissible without affecting adjacent land;
to determine basin morphometry, particularly in areas where water will
leave the basin; and to serve as a base map for planning the locations of
dikes, spillways, water control structures, nesting islands, etc. To survey
a basin, an elevation reading is taken with a surveying transit and leveling rod at enough points in the basin to accurately reflect the contours of
the basin, especially near proposed structures and along property lines
and roads (Figure 3.5). A topographic map is made by plotting the
elevations of all surveyed points and drawing lines of equal elevations.
Appendix D1 provides step-by-step instructions for conducting a simple
basin survey, and Appendix D2 provides a sample data sheet. Some very
simple projects can be surveyed and staked for construction in one field
visit. However, larger or more complicated surveys will often require a
more detailed survey if the basin is selected for restoration. The Minnesota Wetlands Restoration Guide (Wenzel, 1992) provides a clear and
thorough description of construction surveying, developing specifications for contractors, and installing structures. Refer to this manual for
construction surveying for large projects. Usually, a large project should
first be surveyed, then structures should be selected, and finally it should
be resurveyed and staked for earth moving.

The base map should indicate all natural or man-made drainage
features that affect the wetland basin. Many prairie pothole wetlands
were drained initially by excavating a ditch through basins connecting
them to a public drainage system. These ditch systems were often up-

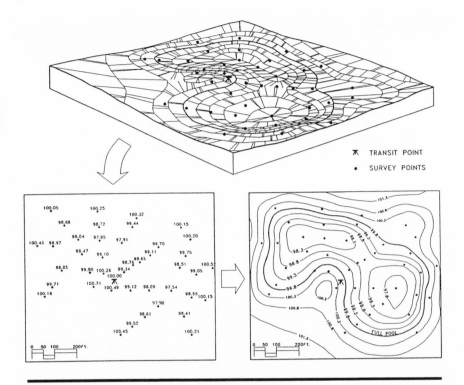

Figure 3.5. The three stages of creating a topographic map of a drained wetland basin in Kandiyohi County, Minnesota.

graded to tile systems, especially in Iowa and southern Minnesota. Some wetlands were originally drained with tile. The location, position, and sizes of the tile lines must be thoroughly investigated to minimize the chance of undetected lines that could continue to drain the site after construction. Usually the most complete information about the drainage system of a site is available from current or past landowners. However, check all possible sources for tile maps: it is easy to overlook or be unaware of old, poorly functioning tiles. Often, the county maintains records on drainage ditches and tiles that are publicly owned and managed by Watershed Districts and Watershed Management Organizations (Minnesota), Drainage Districts (Iowa), or by the county itself. SCS offices usually have copies of maps of these public drainage systems as well as maps of any private drainage that was installed with federal assistance. Draw on a map the locations and diameter of all tiles lines entering and exiting the basin to be restored. The diameter and tile type can be used to calculate the area upslope being drained by this tile.

Tile drains have been installed in five basic patterns: single lines, random systems, parallel systems, double main systems, and herringbone systems (Figure 3.6). In a random system, a main drain lies in the lowest depression in a field and smaller branches that drain smaller depressions feed to the main drain. Random systems are most common in fields with many isolated depressions. Parallel, double main, and herringbone systems have been used to drain larger wetland basins. A single main drain receives flow from either lateral tile lines set perpendicular to one side of the main (parallel) or laterals set at an angle from both sides of the main line (herringbone). Materials used for subsurface drains include clay, concrete, bituminized fiber, metal, and plastic. Most of the tile lines removed during restoration have been clay.

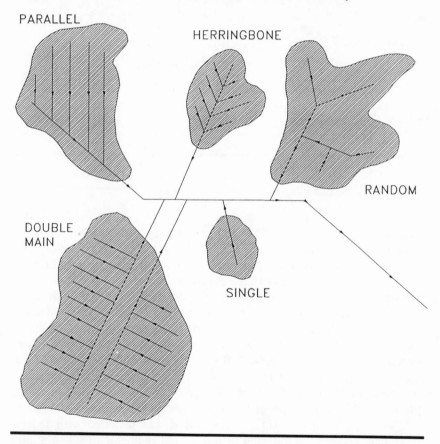

Figure 3.6. Basic tile drainage patterns.

Predicting Water Regime

Besides identifying areas that once were wetlands, the soil series mapped for a basin will suggest its predrainage water regime and vegetation (Figure 3.7). In Appendix C, each depressional wetland soil series in the region has been assigned to a wetland class (assuming the soil type occupies the lowest portion of a basin), using the wetland classification presented in the previous chapter (Stewart and Kantrud, 1971). Except by extrapolation from soils data, it is usually difficult or impossible to determine the type of wetland that formerly occurred in a drained basin because drained basins normally have little or no natural vegetation. Landowners and neighbors may be able to describe the depth and duration of ponding for basins that have been recently drained, but most basins were drained before the present landowner acquired the property. Nevertheless, any data that exist on the water regime of historic wetlands should be sought out to supplement data from soil surveys.

The utility of county soil surveys for restoration planning varies across the region because of differences in scale of the survey and accuracy of the data collected. Many older surveys do not delineate wetland basins and indicate their soil type if they were less than 3 acres. Several of these older surveys have also been found highly inaccurate (Howell, 1990). Appendix C includes a map of the published soil surveys in the region and the status of unpublished field maps. When 66 wetland map units in restored wetlands within the region were field verified, 43 were correctly mapped and 23 were incorrectly mapped (Galatowitsch, 1993). However, only 7 of the mapping errors would have placed the basin in a different potential wetland class. Although these results suggest that soil surveys are likely to be a reliable source of information in the region for planning wetland restorations, local SCS staff should be consulted to determine limitations of the soil survey being used.

Will the historic water regime indicated by soil characteristics of a basin return after tiles are interrupted or ditches blocked? Not necessarily. The restored wetland may not regain its former hydrology because of changes in land use in the watershed, changes in the effective size of the watershed, and regional lowering of groundwater tables. For example, groundwater studies of prairie potholes have shown that water balance in a wetland may be affected not only by on-site conditions (e.g., evapotranspirational losses) but at least as much by local and regional flow systems that depend on the surrounding landscape. Some sites with peaty soils (e.g., Houghton series) often have very small watersheds although their soil suggests a sustained high water table. These sites likely depended on groundwater flow from a relatively large area sur-

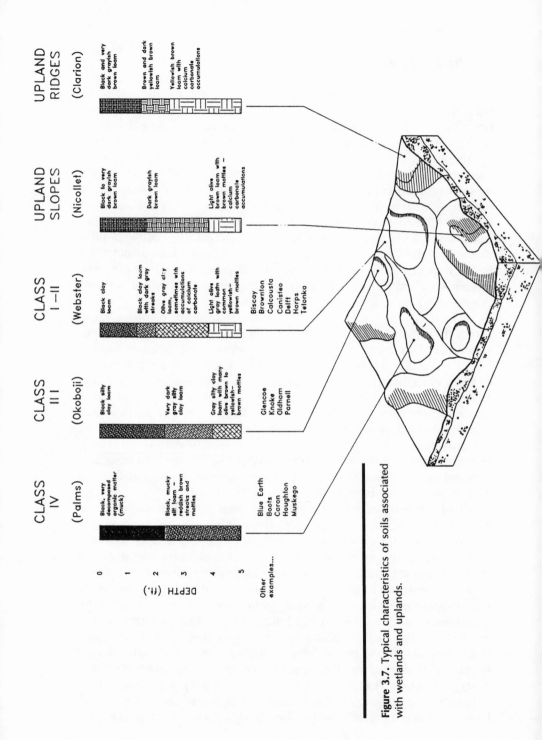

Figure 3.7. Typical characteristics of soils associated with wetlands and uplands.

rounding the basin. Whether historic springs will again flow into these basins is unknown. Regional drainage reduces recharge of groundwater and alters historic groundwater flow patterns (Winter, 1989). On the other hand, some restored basins may receive more water than they did historically because they are supplied by tile lines that drain a larger watershed than the basin had formerly. Drainage ways with soils indicating that a wet prairie or sedge meadow wetland was once present have been restored by blocking tile lines and constructing a dike. These areas, however, became seasonal or even semi-permanent basins because of the increased volume of water entering them.

The nature of the watershed of a restored wetland affects the quantity and quality of water received by the basin. In particular, the watershed's size, elevational relief, and vegetation will affect the volume and rate of runoff. Watershed size can be determined from United States Geological Survey (USGS) 7.5′ quadrangle maps (1:24,000 scale) for basins with large watersheds or for those in hilly areas. First, locate the basin on the map. A soils map may be helpful for finding the basin on the topographic map, especially for small basins. Beginning at the outlet of the basin or the downhill side of the basin (if there is no surface outlet), draw the boundary of the watershed by following the high spots and ridges (Figure 3.8). You can check if the boundary line drawn indicates the edge of a watershed by seeing if areas to either side of your ridge line are lower in elevation. Sometimes it is difficult to find a good ridge line, especially in very flat or hummocky areas. Route your watershed boundary line through the center of broad, flat, high areas. In hummocky areas, other wetlands or drained basins may be within your watershed boundary. This is reasonable if the other basins are at higher elevations than the project basin.

Watersheds of basins that are small or have flat terrain will often be difficult to delineate on USGS topographic maps because of inadequate elevation detail. In these cases, use soils maps to delineate watersheds. The same principles apply, but instead of following ridge lines identified by contour lines, route the watershed boundary line through the center of soil types typical of ridges (for example, *Clarion* series). Use your published county soil survey to identify the ridge soils near your project. This information, often with useful diagrams, is generally contained in the opening descriptive chapters of the soil survey.

As you establish the boundary of the watershed, be sure to include the land that would be drained by all surface inlets shown on the map. Be especially careful with areas where drainage is intercepted by roads and field accesses and routed through culverts. Look for the location of culverts in the watershed area during the field visit if the restoration

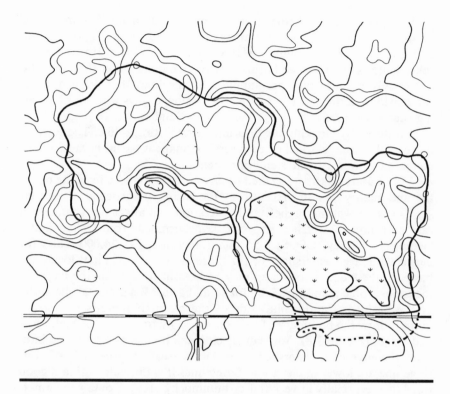

Figure 3.8. The watershed of Four Mile Marsh, a restored wetland in Emmet County, Iowa (T99N R34W S 08). Map from a 7.5-minute USGS quadrangle. Contour interval is 5 feet.

project is near roads or other obstructions. Storm runoff through culverts can be very high and needs to be considered when designing the outlet structures for the project. Sometimes, tile lines will connect basins and low-lying areas together that would otherwise have not been integrated. In these cases, the watershed of restored basins may be larger than that of its surface watershed. In other cases, the artificial drainage system is entirely within the natural watershed. Draw watershed lines that include all areas within the tile drainage area feeding a restored basin.

Calculate the size of the basin and the watershed delineated on your map. The easiest way to make these calculations is to overlay a dot area counter on the map, count the dots, and find the acreage associated with the dot count (Figure 3.9). Figure 3.10 can then be used to determine if the basin will be flooded through midsummer in most or all years, some

		Example:
1. How many dots are in each acre on your map? Place a dot grid over the map. Count the number of dots in any quarter section (160 acre) area on your map. Your conversion factor is this dot count divided by 160.		Number of dots in quarter section = 320 Conversion Factor (C) = 320/160 = 2
2. How many dots are included in the watershed and in the basin? Place the dot grid over the watershed area and count the dots across the entire area (including the basin). Then, count the dots in the basin alone.	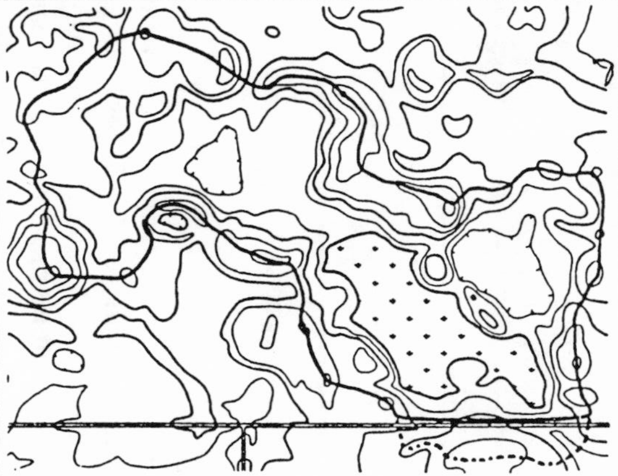	From map at right: Dots counted in watershed = 125 Dots counted in basin = 30
3. What is the size of the watershed and basin? Multiply the dot counts for the watershed and basin by the conversion factor.		Watershed Size: 125 dots x 2 dots per acre = 250 acres Basin Size: 30 dots x 2 dots per acre = 60 acres

Figure 3.9. Calculating basin and watershed size using a dot grid.

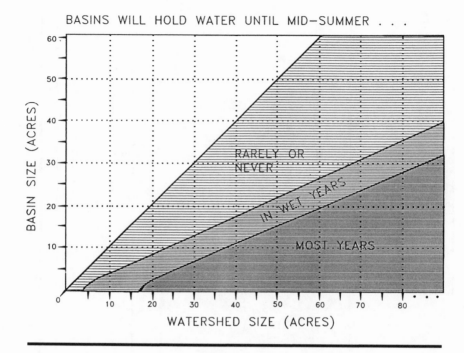

Figure 3.10. The predicted water regimes of restored wetlands on the basis of their watershed size.

years, or never. This graph was generated from water depth records collected by landowners and managers for 62 wetland restorations across the region (Galatowitsch, 1993). Monitoring of basins within the region showed that water was present in midsummer most years in restorations with watersheds of 17 acres or more and at least 4 acres of watershed for each acre of basin. Wetlands with less than 2.5 acres of watershed for each basin will rarely have standing water. No matter what the basin size, wetlands with less than 4 acres of watershed will also rarely have standing water. Restored wetlands with intermediate characteristics will likely be seasonal wetlands, holding water until midsummer only in wet years.

Revegetation Potential

There are two possible ways that a restored basin will become revegetated in the short term. Wetland species may have persisted in the basin after drainage or in drainage ditches. Seeds of wetland species may still be present in the seed bank. The spacing of tiles and ditches, the age and maintenance of drainage systems, and the total load on downstream drains affected how effectively a basin was drained and, consequently, if wetland species could or could not persist in the basin. Many wetland plants, such as three square bulrush and smartweeds, will grow in cultivated areas that ponded periodically. Whether or not seeds of wetland species are still present in the seed bank depends on both how long the basin has been drained and how effectively it was drained. The potential for natural revegetation varies from basin to basin. There is little potential for a basin that was effectively drained and rarely ponded while one that was poorly drained and had standing water several times a year has a high potential. Consequently, any information that can be obtained from the current landowner, past landowner, or neighbors on the duration and effectiveness of drainage and if any wetland plants have been seen in the basin while it was under cultivation will be useful for evaluating revegetation potential. Appendix D3 includes a sample data collection form for recording site history. Appendix D4 is a data sheet for recording preconstruction conditions, including soils and vegetation.

For wetlands drained effectively, the best candidates for natural revegetation are those that have been drained less than 20 years because their seed banks will still contain viable seeds of many wetland species. However, most restoration sites in the region have been drained for more than 20 years—in fact many have been drained for 50 to 80 years. It may be desirable to examine the composition of the seed bank, especially for large costly projects. Appendix D5 outlines how to do seed bank studies.

Dispersal of new species to the site after restoration will undoubtedly also occur, but little is known about what species and when this will occur. A study of created wetlands in southeastern Wisconsin showed that as distance to nearest seed source increased, the number of native wetland species decreased (Reinartz and Warne, 1993). More information is needed before we can generalize about how close restorations must be to seed sources to encourage revegetation. It seems reasonable to assume that restorations near or connected to natural wetlands will have greater supplies of propagules.

A plant list should be compiled as part of a preconstruction site assessment. Remnant wetland species in the basin will often expand rapidly and become the dominant vegetation in the restored wetland. Sites with undesirable plant species, such as reed canary grass or purple loosestrife, should be given a lower priority for restoration than sites without these species. Surviving wetlands within the drained basin should also be identified during the preconstruction plant survey. For example, an impoundment was planned in northern Iowa that would have flooded a nearby fen harboring rare plants. The proposed project included a high dike that would have increased the water level in the basin to above that of the elevation of the fen. Because the fen was recognized during the field inspection and the potential for flooding also was recognized, the project was modified by lowering the elevation of the proposed dike.

What plant species will potentially become established in a basin is predictable from expected or actual hydrology; that is, will it flood into midsummer most or all years or only in wet years (Figure 3.10)? Those that flood into midsummer most years will likely become seasonal or semi-permanent marshes (Class III and IV), whereas those that do not will likely become ephemeral or temporary marshes (Class I and II). Lists of species associated with different wetland classes are found in Chapter 2, Appendix A, and in Kantrud et al. (1989). There is no certainty that these species will actually occur in a particular restored wetland in the short term (see Chapter 5). It is possible, but far from certain, that the expected species may colonize a particular basin in the long term. Whether in the long term means after 10, 100, or 1,000 years is unknown. Restorations being done to establish wildlife habitat should give priority to basins in which natural revegetation is most likely to occur.

The widths of vegetation zones in restored wetlands will be largely a function of the slopes of the basin. Broad zones will result if there are gentle slopes, narrow zones from steep-sided basins. In Figure 3.11 the changes in water surface area in two basins are compared that have similar basin and watershed characteristics but differ in basin morphometry. Basin B is actually a U-shaped impounded ravine whereas Basin A is more typical of a shallow prairie pothole with gently sloped sides. Basin B will not develop broad bands of sedge meadow and emergent vegetation because water depth rapidly increases from the edge of the marsh, beyond the tolerance of aquatic plants. In contrast, a rather broad sedge meadow and emergent rim could develop around Basin B.

PERMITS

Restoring wetlands in agricultural areas requires consideration of laws designed to ensure that agriculture on adjacent lands will not be adversely affected. Restorations that modify a tile line must be designed so flows from upstream properties into the basin are not impeded. Downstream owners do not have the right to send water upstream, so restoration projects must not flood surface waters back onto adjacent landowners without their explicit permission.

Besides respecting rights of adjacent landowners, restoring wetlands requires consideration of local, state, and federal regulations that govern water use and protection, dam construction, and public drainage. Government agencies often enforce these regulations through issuance of

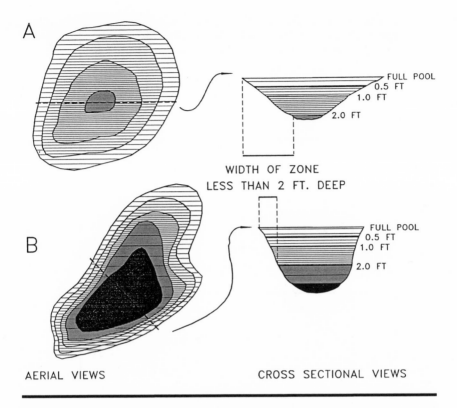

A

FULL POOL
0.5 FT
1.0 FT
2.0 FT

WIDTH OF ZONE
LESS THAN 2 FT. DEEP

B

FULL POOL
0.5 FT
1.0 FT
2.0 FT

AERIAL VIEWS   CROSS SECTIONAL VIEWS

Figure 3.11. Shallow depressions (A) usually will have broad vegetation zones whereas steep-sided basins (B) will often have narrow vegetation zones.

permits. One to four permits may be required for a typical wetland restoration, depending on its size, location, and potential adverse impacts. In general, restorations of large basins will likely require more permits than smaller projects. Some agencies require permit application fees and rather extensive documentation of restoration plans whereas others may not. Assistance in completing required application materials is often available from local representatives of government agencies or from conservation organizations such as Ducks Unlimited. Sometimes, state and federal managers may already have the permits needed to restore many wetlands within their area, eliminating the need for individual permits. However, care should be taken to ensure all required permits are obtained; the penalties for not doing so are sometimes severe. A summary of required permits is provided in Table 3.3. Since regulations frequently change, contact agencies listed to obtain current application information.

TABLE 3.3. Permits that may be required for wetland restorations. (Contact agencies for current application information.)

PERMITS REQUIRED IN...	NAME OF PERMIT	ACTIVITIES REGULATED	PERMITS REQUIRED FOR ...	INFORMATION REQUIRED (IN ADDITION TO APPLICATION FORMS)	CONTACT THE FOLLOWING AGENCIES FOR APPLICATION FORMS AND SUBMISSION GUIDELINES
ALL STATES (FEDERAL PERMIT)	Section 404 Permit (often a Nationwide Permit for wetland restorations)	Placement of dredge or fill material (such as soil) into wetlands, for instance from dam or nesting island construction..	All wetlands.	Initially: ◆ Brief description of the project; ◆ Drawing showing proposed work; ◆ Photographs of the site, if available. The U.S. Army Corps of Engineers will review the completed application and preliminary information to determine if more detailed information is required.	Iowa: U.S. Army Corps of Engineers Rock Island District Clock Tower Building P.O. 2004 Rock Island, IL 61204 Phone: 309-788-6361 Minnesota: U.S. Army Corps of Engineers St. Paul District 180 Kellogg Blvd. East Room 1421 St. Paul, MN 55101 Phone: 612 - 220 - 0200 South Dakota: U.S. Army Corps of Engineers Omaha District 215 No. 17th St. Omaha, NE 68102 Phone: 605 - 221 -4211

66

TABLE 3.3. (*Continued*)

PERMITS REQUIRED IN...	NAME OF PERMIT	ACTIVITIES REGULATED	PERMITS REQUIRED FOR	INFORMATION REQUIRED (IN ADDITION TO APPLICATION FORMS)	CONTACT THE FOLLOWING AGENCIES FOR APPLICATION FORMS AND SUBMISSION GUIDELINES
ALL STATES (LOCAL PERMIT)	Drainage Agreement	Modification of public (also called legal) tile lines or ditches.	All wetlands directly drained by a public tile line or ditches.	Variable, but may include: ◆ Description of the project including the size and location of tile or ditch and its connections; ◆ Description of the current condition of the drain system and the natural drainage flow; ◆ Cross-sectional map of the proposed construction as related to the original drainage profile; ◆ List of responsible parties for engineering, construction and costs of the project.	If the area is within an organized watershed or drainage district, contact them for guidelines on submitting information for approval. If the basin is not in a watershed or drainage district, contact the county board of commissioners or supervisors for this information. Phone: see your phone directory under government listing.
IOWA	Floodplain Development Permit	Construction, operation, and maintenance of dams.	All wetlands constructed with an embankment of more than five feet and that store more than 18 acre feet of water.	◆ The prescribed order of work, and names of persons responsible for project design and construction. ◆ Technical provisions which describe work methods, equipment, materials, and desired end results. ◆ Map indicating ownership and land use of areas to be occupied by the dam embankment, spillways, other structures; areas within the maximum normal pool; areas that could be affected by temporary flooding or spillway discharge.	Iowa Department of Natural Resources Water Resources Section Wallace State Office Building 900 East Grand Avenue Des Moines, IA Phone: 515- 281- 5029

TABLE 3.3. (*Continued*)

PERMITS REQUIRED IN...	NAME OF PERMIT	ACTIVITIES REGULATED	PERMITS REQUIRED FOR	INFORMATION REQUIRED (IN ADDITION TO APPLICATION FORMS)	CONTACT THE FOLLOWING AGENCIES FOR APPLICATION FORMS AND SUBMISSION GUIDELINES
MINNESOTA	Water Permit (Includes dam safety, water quality, and archeological review)	Modifications to the current, course, or cross-section of Minnesota protected waters or wetlands, including construction of dams, installation of culverts..	Lakes, natural water courses with a watershed over 2 square miles, OR wetlands over 10 acres (2.5 acres in municipalities). Maps of protected waters and wetlands are available for inspection at county SCS and regional DNR offices.	◆ Plans describing the proposed project; ◆ Brief statement of the anticipated changes in the water and related land resources, detrimental effects, and alternatives to the proposed actions; ◆ Proof that appropriate local government agencies have been notified of the proposed project.	Minnesota Department of Natural Resources Division of Waters DNR Building, Third Floor 500 Lafayette Road St. Paul, MN 55155 Phone: 612 - 296 - 4800
SOUTH DAKOTA	Water Permit	Modifications or diversions of the normal flow of surface or ground water.	Wetlands constructed that dam dry draws and impound more than 25 acre-ft of water; OR wetlands that dam watercourses OR wetlands that include physical diversions, such as for irrigation..	◆ Map showing impoundment in relation to a government section corner, the location of the high water line and lands under water (including ownership), and capacity of maximum pool. ◆ Brief description of the proposed project. ◆ Cross-sectional diagram of dam.	State of South Dakota Water Rights Division Dept. of Environment and Natural Resources Joe Foss Building Pierre, SD 57501 Phone: 605 - 773 -3352

REFERENCES

Bartels, R.M. 1993. Personal communication, U.S. Soil Conservation Service Conference on Wetland Restoration, Fergus Falls, Minnesota.

Bowles, J.B. 1975. *Distribution and Biogeography of Mammals of Iowa.* Special publication of the museum, Texas Tech University, Lubbock No. 9. 184 pp.

Brown, M. and J.J. Dinsmore. 1986. Implications of marsh size and isolation for marsh bird management. Journal of Wildlife Management 50: 392–397.

Christiansen, J.L. and R.M. Bailey. 1988. *The Lizards and Turtles of Iowa.* Iowa Department of Natural Resources. Nongame Technical Series No. 3. 19 pp.

Christiansen, J.L. and R.M. Bailey. 1990. *The Snakes of Iowa.* Iowa Department of Natural Resources. Nongame Technical Series No. 1. 16 pp.

Christiansen, J.L. and R.M. Bailey. 1991. *The Salamanders and Frogs of Iowa.* Iowa Department of Natural Resources. Nongame Technical Series No. 3. 24 pp.

Conant, R. and J.T. Collins. 1991. *A Field Guide to Reptiles and Amphibians in Eastern and Central North America.* Third Edition. Peterson Field Guide Series. Houghton Mifflin Company, Boston.

Crumpton, W.G. 1989. Algae in northern prairie wetlands. IN A.G. van der Valk (Ed.). *Northern Prairie Wetlands.* Iowa State University Press, Ames.

Dinsmore, J.J., T.H. Kent, D. Koenig, P.C. Peterson, and D.M. Roosa. 1984. *Iowa Birds.* Iowa State University Press, Ames.

Galatowitsch, S.M. 1993. Site selection, design criteria, and performance assessments for wetland restorations in the prairie pothole region. Ph.D. Dissertation, Iowa State University, Ames.

Hamilton, N.D. 1990. *What Farmers Need to Know about Environmental Law.* Iowa Edition. Drake University Law Center, Des Moines, Iowa. 189 pp.

Hazard, E.B. 1982. *The Mammals of Minnesota.* University of Minnesota Press, Minneapolis. 280 pp.

Howell, K. 1990. Personal communication, U.S. Soil Conservation Service, Brookings County, South Dakota.

Isenhart, T.M. 1992. Transformation and fate of non-point source nitrate loads in northern prairie wetlands. Ph.D. Dissertation, Iowa State University, Ames.

Jacques, J.E. and D.L. Lorenz. 1987. Techniques for estimating the magnitude and frequency of floods in Minnesota. U.S. Geological Survey. Water Resources Investigations Report 87-4170.

Janssen, R.B. 1987. *Birds in Minnesota.* University of Minnesota Press and James Ford Bell Museum of Natural History, Minneapolis. 352 pp.

Jones, J.K., Jr., D.M. Armstrong, and J.R. Choate. 1985. *Guide to Mammals of the Plains States.* University of Nebraska Press, Lincoln. 371 pp.

Jones, J.K., Jr., D.M. Armstrong, R.S. Hoffman, and C. Jones. 1983. *Mammals of the Northern Great Plains.* University of Nebraska Press, Lincoln. 378 pp.

Kantrud, H.A., J.B. Millar, and A.G. van der Valk. 1989. Vegetation of wetlands of the prairie pothole region. IN A.G. van der Valk (Ed.). *Northern Prairie Wetlands.* Iowa State University Press, Ames.

Moore, I.D. and C.L. Larson. 1980. Hydrologic impact of draining small depressional watersheds. Journal of the Irrigation and Drainage Division, Proceedings of the American Society of Civil Engineers 106(IR4): 345–363.

Neely, R.K. and J.L. Baker. 1989. Nitrogen and phosphorus dynamics and the fate of agricultural runoff. IN A.G. van der Valk (Ed.). *Northern Prairie Wetlands*. Iowa State University Press, Ames.

Omernik, J.M. 1977. Nonpoint source-stream nutrient level relationships: a nationwide study. Environmental Protection Agency, Corvallis, Oregon. 150 pp.

Reinartz, J.A. and E.L. Warne. 1993. Development of vegetation in small created wetlands in southeastern Wisconsin. Wetlands 13(3): 153–164.

South Dakota Ornithologists' Union. 1991. *The Birds of South Dakota*. Second Edition. Northern State University Press, Aberdeen, South Dakota. 411 pp.

Stewart, R.E. and H.A. Kantrud. 1971. Classification of natural ponds and lakes in the glaciated prairie region. Research Publication 92. Bureau of Sport Fisheries and Wildlife.

Thompson, M.L. 1987. Micromorphology of four Argialbolls in Iowa. IN N. Fedoroff (Ed.). *Proceedings of the Seventh International Working Group on Soil Micromorphology* (Paris). Association Française pour l'Étude du Sol, Plaisir, France.

Thompson, M.L. 1992. Personal communication, agronomy department, Iowa State University, Ames.

van der Valk, A.G. and R.W. Jolly. 1992. Recommendations for research to develop guidelines for the use of wetlands to control rural nonpoint source pollution. Ecological Engineering 1: 115–134.

Wenzel, T. 1992. *Minnesota Wetlands Restoration Guide*. Minnesota Board of Water and Soil Resources. St. Paul, Minnesota.

Winter, T.C. 1989. Hydrologic studies of wetlands in the northern prairie. IN A.G. van der Valk (Ed.). *Northern Prairie Wetlands*. Iowa State University Press, Ames.

DESIGN CONSIDERATIONS

I n the 1950s when constructing impoundments for wildlife first became popular, a "standard design" was promoted for such impoundments (Atlantic Flyway Waterfowl Council, 1959). The standard impoundment became a small reservoir created by putting a dike across a ravine or draw. The adoption of a standard design undoubtedly resulted in more impoundments being constructed than would have been done otherwise, but it also resulted, of course, in large numbers of nearly identical impoundments being built. Even today basins are sometimes targeted for restorations in the prairie pothole region because they are well suited to this standard design. We are not in favor of a standard design for restoring prairie potholes. As outlined in the last chapter, general guidelines for restoring prairie potholes can be either historical or functional. In either case, restoration designs need to be developed that are suitable for the type of wetlands being restored and the function(s) for which they are being restored. A restoration project to establish a temporary wetland for rail and bittern habitat will need a different design than a project to establish a semi-permanent wetland for dabbling ducks. The former may require only a tile break, whereas the latter may require a dike to retain additional water in the spring and a water control structure to permit water-level manipulation.

Restoration designs are primarily concerned with site modifications that will be needed to create the desired hydrology in the restored wetland. These include plugging ditches or removing tiles, building dikes, and excavating the basin. The specific modifications needed in a basin will be dependent both on the nature of the basin and its watershed, as determined by the restoration feasibility assessment of the site, and the goals of the project (e.g., water quality improvement). Guidelines for developing detailed construction plans and specifications for such modifications is beyond the scope of this book. The *Minnesota Wetland Restoration Guide* (Wenzel, 1992) and U.S. Soil Conservation Service (SCS) Engineering Field Manual (1991) both provide information on developing construction plans and specifications, and these manuals should be consulted for this purpose. In this chapter, we will discuss five essential topics and, when relevant, their functional implications that should be considered when designing prairie pothole restorations: plugging ditches and removing drainage tiles, dike construction, water control structures, basin excavation, and watershed land use.

A checklist and worksheet are given in Appendix D7 for determining what site modifications may be needed and for collecting and organizing data needed for the design of these modifications.

PLUGGING DITCHES AND REMOVING TILE LINES

Wetlands can sometime be restored simply by plugging ditch lines at their outlet (Figure 4.1). This is feasible when basins are drained with shallow ditches (less than a few feet in depth). However, ditch plugs have not been successful for restoring basins drained with deep ditches be-

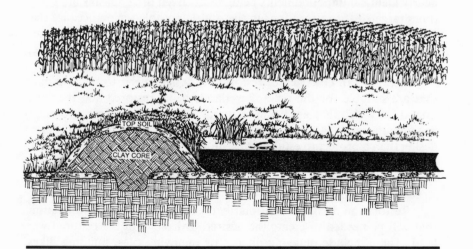

Figure 4.1. A ditch plug used to close surface drainage systems.

cause deep ditches act as sumps and continue to lower the water table in restored basins. Sometimes ditch lines may be deep enough to penetrate the natural seal of basins, permitting water to flow into permeable materials beneath them. These ditches will need to be sealed before being filled. In short, deep ditches need to be filled to grade in order to restore a basin's hydrology.

In basins drained with tile, all tile lines entering the basin must be located and either they must be broken or a section must be removed and replaced with impermeable tile. When landowners or managers wish to retain the option of reconnecting the tile, a section of tile should be removed at the outlet (Figure 4.2). The length of the tile removed should be a minimum of 50 feet in heavy clay and 150 feet in sandy and organic soils. These distances should be adequate to prevent seepage into still functioning tile. If tile lines are broken, the ends of the tile in the basin are often capped to prevent sediments from entering it. Nonpermeable replacement tile should always be at least as large as the original drainage tile to ensure adequate flow through the basin and to prevent tile ruptures in situations where tile lines need to remain in working condition. A riser is normally installed at the end of the input tile to bring water to the surface, and another riser on the output tile to control water levels in the basin (Figure 4.2) if no other water-level control structures are installed.

For projects designed to improve water quality, input and output risers should be as far apart as possible (Figure 4.3). This ensures that water has to flow the entire length of the basin and increases its potential residence time in the wetland. The longer the residence time, the greater

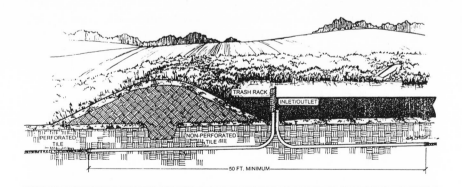

Figure 4.2. A tile break created by replacing a section of drainage tile with impermeable tile. Inlet and outlet risers are side by side.

74

CHAPTER 4

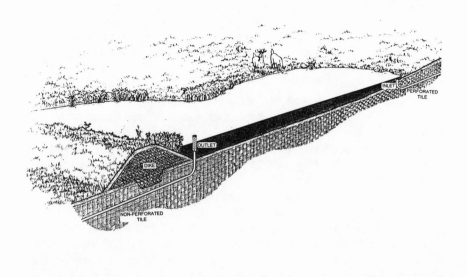

Figure 4.3. A tile break with the inlet and outlet separated to increase water residence time in the basins. This should increase the effectiveness of this wetland for improving water quality.

the opportunity for nutrient and contaminant removal. The effect of increased residence time on nitrate levels is evident from observations of two natural wetlands in northern Iowa (Figure 4.4). The difference in the effectiveness of these wetlands for removing nitrates seems to be largely a function of the proximity of their inlets and outlets. Likewise, in wetlands drained with ditches, water often flows through the wetland along these old channels and can reduce residence time. Grading the basin during construction to fill ditches should prevent such channel flow within a restored wetland.

All subsurface and surface drains within a basin must be interrupted to ensure a successful restoration. Occasionally, tile lines are undetected during construction and the basin does not retain water as anticipated. In at least one instance, wetland managers compromised with a landowner and removed one functioning tile but allowed a second to remain. This basin has not and will likely never retain any water.

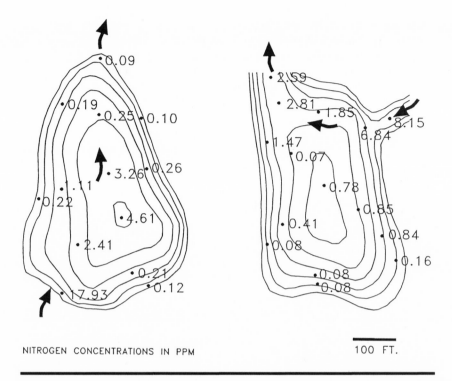

NITROGEN CONCENTRATIONS IN PPM

100 FT.

Figure 4.4. The increased travel distance of water flowing through the basin at the left results in very low nitrate concentrations in the outlet whereas in the similar basin on the right, which has a shorter flow path between the input and output, nitrate concentrations are much higher in the outlet. Both sites are natural wetlands located in Palo Alto County, Iowa, that receive surface runoff from cornfields (right basin: T97N R34W S23; left basin: T96N R33W S31).

DIKE CONSTRUCTION

The most common modification needed is a dike to raise water levels and/or to prevent flooding of adjacent land, roads, utility corridors, etc. How often the basin floods to design capacity, or if it ever will, is to a large extent a function of the size of the watershed. Dikes to raise water level should only be considered if the restored basin will receive sufficient water to fill the basin to full design capacity and only if dikes are needed to achieve water depths required by the goals of the project. The placement and height of the dike that is needed to restore a basin so that it functions as planned can be determined using the topographic map of the basin that was constructed as part of the site evaluation (see Chapter 3). The elevation at which the basin will be full and water will begin to flow through outlets or over the spillway is called full pool elevation. As mentioned previously, the elevation at which full pool elevation should be set will sometimes be determined by property lines while other times it will be determined solely by project goals.

What effect different full pool elevations will have on the amount of the basin flooded to different water depths can be estimated by constructing a table of the areas between contour intervals and, for each potential full pool elevation, determining the number of acres at different water depths (see Table 4.1). The optimal full pool elevation depends on the goal of the project. For example, if the optimal water depth for the project whose basin morphometry is summarized in Table 4.1 is 1–3 feet, the maximum area (30 acres) between these water depths for this basin would occur with a full pool elevation of 110 feet. Most restored prairie potholes will not remain at full pool throughout the year, and this should be taken into account for projects whose primary function is as wildlife habitat.

Portions of the original basin may not be restorable because of land ownership boundaries or obstructions such roads, railroad beds, or cropland (Figure 4.5). By necessity, restorations that cannot flood across the entire basin will require retaining structures to confine the flooded area. An earthen dike incorporating a principal spillway and an emergency spillway is the common retaining structure used in prairie pothole restorations. Occasionally such dikes have been constructed through the center of basins. In some cases, water ponded on both sides of the dike (Figure 4.6) created serious problems on the "nonrestored" section of the basin. Restorations that include the entire basin are usually going to be

TABLE 4.1.   Contour areas and water depth (ft) at four potential normal pool elevations for a basin to be restored

CONTOUR INTERVAL (ft)	AREA (acres)	NORMAL POOL ELEVATION (FT)			
		108 ft max	109 ft max	110 ft max	111 ft max
103-104	1	4-5	5-6	6-7	7-8
104-105	3	3-4	4-5	5-6	6-7
105-106	6	2-3	3-4	4-5	5-6
106-107	8	1-2	2-3	3-4	4-5
107-108	16	0-1	1-2	2-3	3-4
108-109	14		0-1	1-2	2-3
109-110	7			0-1	1-2
110-111	7				0-1
TOTAL AREA	(acres)	34	48	55	62

Source: Anderson, 1985.

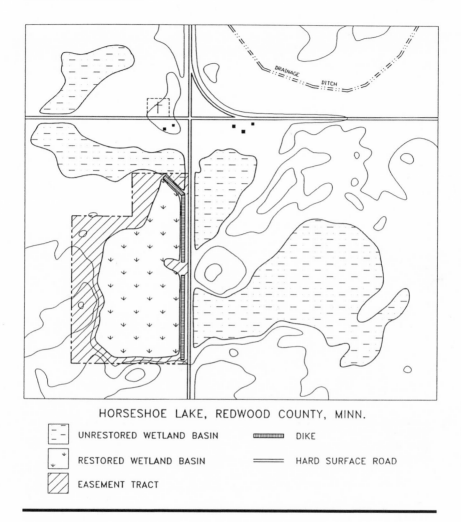

HORSESHOE LAKE, REDWOOD COUNTY, MINN.

- UNRESTORED WETLAND BASIN   🔲🔲🔲 DIKE

- RESTORED WETLAND BASIN   ═══ HARD SURFACE ROAD

- EASEMENT TRACT

Figure 4.5. Design of a project in which only a portion of this wetland basin could be restored in Redwood County, Minnesota (T111N R39W S29), because the remaining portions of the drained basin are on other properties and crossed by a road. A dike was constructed parallel to the roadbed to retain water within the easement tract.

simpler to design, cheaper to construct, and less trouble to maintain because retaining dikes are not needed. They are also more likely to be successful because the basin's watershed has not been divided. Consequently, basins that can be restored in their entirety should be given preference over basins that can only be partly restored.

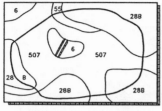

PORTION OF SOIL SURVEY
OF KOSSUTH COUNTY, IOWA

SOIL   UNITS

6	Okoboji silty clay loam
507	Canisteo clay loam
55	Nicollet loam
28B	Dickman fine sandy loam, 2–5 % slopes

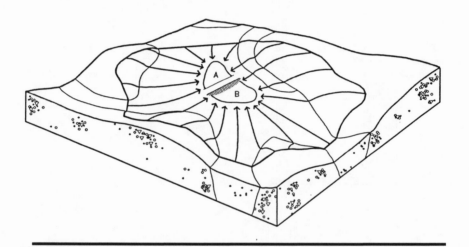

Figure 4.6. A dike was constructed through the middle of this basin in Kossuth County, Iowa. Although not intended, water ponded on both sides of the dike.

Some general guidelines are available for the construction of dikes. These dike designs are suitable for agricultural areas where damage from dike failure is minimal and the maximum water depth against the dike is 6 feet or less for mineral soils and 4 feet or less for organic soils. Dikes for larger projects or projects in high-risk locations should be designed by a qualified engineer. All dikes need to be properly sited, designed, and constructed to avoid structural failure. Overtopping during high flow, undermining from channel flow, sloughing from wave action or saturation of unstable materials, piping or excavation from burrowing animals, and seepage along water control structures placed through embankments are common causes of dike failure.

Most smaller basins in the prairie pothole region have a clay subsoil that is suitable dike material. Dikes should normally not be constructed from organic soil (top soil or peat) because excessive settlement, shrinkage, and sloughing are likely. The foundation area for the dike should be cleaned of trees, stumps, logs, roots, brush, and boulders. A core trench should be excavated from this area, removing the organic matter to a width of 4 feet. Some basins may have a sandy or even gravelly subsoil if near an old stream channel. Likewise, reaching impermeable subsoil may be difficult in large basins with deep peat deposits. If the subsoil is permeable, a deeper core trench must be dug and filled with clay or bentonite before the embankment is created. A diaphragm cutoff wall can be embedded in the dike if the best available core material will likely allow unacceptable amounts of seepage. The cutoff wall is constructed by driving sheet steel or wood piling into an impervious foundation.

Side slopes of the dike should be no steeper than 3:1. Increasing the side slope increases the stability of a dike. The top width of an embankment should be 6 feet for dikes under 6 feet in height and 8 feet for dikes between 6 feet and 10 feet high. The top elevation of the embankment should be at least 2 feet above normal pool elevation. A lining of hardware cloth or chicken wire will reduce damage from burrowing animals such as muskrats. These liner screens can be positioned vertically in the dike but are probably more effective if laid over the upstream slopes of the dikes and covered with at least a foot of fill.

The dike will often settle after construction because of compaction of the fill material and decomposition of organic matter. Allowances need to be considered during construction so the dike will be adequate to retain the anticipated pool levels. Settlement allowances depend on the soil materials in the fill and foundation, and on the method of construction and moisture content of soil during construction. The following increases in the amount of fill needed to offset settling are recommended by the SCS: 5% when dike is compacted fill, 10% when fill is dumped and shaped, 20% for dragline construction, and 40% for dikes constructed of fill high in organic matter. For example, if the design height of a dike is 8 feet and the fill material is compacted mineral soil, the dike should be overbuilt by 0.4 feet to ensure it will be at the proper height after settling. However, if the design height of the dike is 8 feet and the fill material is organic material, 40% more fill is needed; i.e., the dike should be overbuilt 3.2 feet. If organic matter must be used over the core trench, it is advisable to increase the embankment width and then cover the dike with a clay coating 6–12 inches deep to reduce oxidation or potential decomposition of organic soil and consequent settling. Fill material should be built up on the embankment in 8- to 10-inch layers (loose depth). Compact each layer by passing over the area at least three times.

For impoundments with dikes 1 foot or less in height, a vegetated spillway is adequate. Larger projects will require both a primary spillway through the embankment and an emergency spillway (vegetated). The primary spillway or outlet should have the capacity to pass peak flow expected once in 10 years (as measured over a 24-hour period) with the water level returning to the prestorm elevation within 24 hours. Pipes need to be greater than 6 inches in diameter and risers greater than 8 inches to prevent clogging with sediment and debris. Straight pipe and drop inlet spillways are typically installed in prairie pothole restorations. In both cases, a pipe is buried in the dike. If a tile line has been interrupted as part of the restoration, this pipe is often attached to the tile leaving the basin. A straight pipe spillway does not have an outlet riser. The elevation of the pipe in the embankment sets maximum pool elevation (Figure 4.7). Consequently, water level cannot be manipulated on a straight pipe spillway. For a drop inlet spillway, the elevation of a flexible outflow pipe or a flexible or rotatable extension of the pipe can be adjusted to raise or lower the water level in the basin (similar to Figure 4.2). Seepage, piping, and erosion around discharge tubes should be controlled by using a drainage diaphragm or antiseep collars. The drainage diaphragm is a sheet metal flange on the spillway pipe. A more elaborate water control structure can also be installed to control water levels in larger projects.

The riser on the primary outlet riser is usually perforated to allow

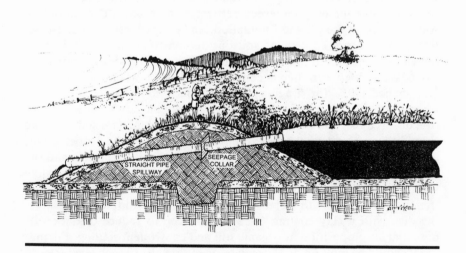

Figure 4.7. A straight pipe outlet with a seepage collar.

for gradual drawdown in water levels after storm events. The "crest" of the riser should be at least a foot below the height of the emergency spillway. Perforations are 1-inch holes at 8-inch intervals vertically and 10–12-inch intervals horizontally. Antivortex trash racks to maintain unobstructed flows should be placed over the top opening of the risers. Perhaps the most common problem in managing restored wetlands is maintaining outflow through spillway pipes. Debris clogs the perforations, and often even the trash rack, and can cause water levels to rise to the level of the emergency spillway. All risers must be cleaned frequently, especially during and after major runoff events. Trash exclosures have been installed in some projects to try to avoid debris clogging the outflow pipe. Usually these are fences built around the outlet.

Earthen, vegetated channels are often built on one side of the embankment to discharge flows greater than the principal spillway can handle or to be the only overflow areas on small dikes. This is the emergency spillway. Earthen spillways should be protected against erosion by seeding or sodding. Mulching may be necessary to protect seedlings.

WATER CONTROL STRUCTURES

Large marshes restored to establish wildlife habitat should be equipped with water control structures to allow drawdowns, especially when dikes are used to impound water. Water-level structures are needed to drain these wetlands periodically in order to reestablish emergent species and restore them to the hemi-marsh condition (Bishop et al., 1979). Even wetlands that are expected to be flooded nearly all years should be designed to allow periodic drawdowns so that water levels can be lowered to allow the repair of structures or dikes. Water control structures may be associated with the primary spillway of a dike, although they need not be. Where a section of drainage tile has been removed and replaced with impermeable tile, a *whistletube* (Wisconsin tube) structure can be installed along the impermeable tile section (Figure 4.8). This is generally a metal, fiberglass, or steel box or a culvert that is divided into two halves with stop logs. A fiberglass version (DOS-IR valve) is frequently used in small projects (Figure 4.9). The top of the structure should extend 1 foot or 2 above normal pool level. A variety of water control valves, such as those used in a city municipal water system, have also been used successfully in prairie pothole restorations, but these can be quite expensive.

Figure 4.8. A Wisconsin tube water control structure along a tile break.

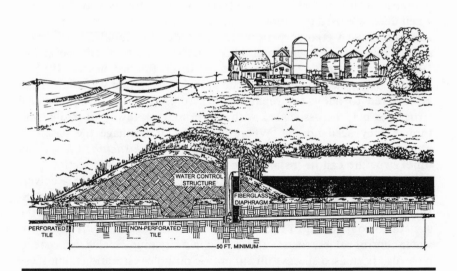

Figure 4.9. A DOS-IR valve water control structure along a tile break.

BASIN EXCAVATIONS

Increasing the water depth of a restored basin is often essential to create better waterfowl habitat. There are two ways to increase the maximum water depth of a basin: (1) install a dike with a spillway that raises the water level to a greater elevation than would have existed naturally, and (2) excavate a section of the basin to the desired depth. Because increasing the full pool elevation will also increase the extent of the basin, this option may not be feasible where potential flooding of adjacent property could occur. Increasing water depth by excavating the basin is limited by the distance to permeable subsoil — such as sand and gravel — that may lie beneath the wetland. In addition, excavations are often very costly. Frequently, water depth is increased by a combination of excavation and dike construction by taking material for the dike construction from within the basin.

Creating a permanent hemi-marsh is the goal of most excavations by deepening some areas beyond the water depth tolerance (about 3 feet) of emergent species. Excavations in small restorations, however, often have unintended negative consequences for many species of waterfowl. Excavating areas in small wetlands reduces the area available for foraging for dabbling ducks, the target species for which these wetlands are usually restored. In Oklahoma, waterfowl select natural shallow wetlands over impoundments that have mostly deep open water (Fredrickson, 1985). Deepened wetlands will not be utilized by spring-migrating ducks because they do not thaw early enough. Excavations that create steep slopes in a basin can increase waterfowl predation because waterfowl concentrate to feed in shallow water near shorelines where they are more accessible to foxes (Olson, 1990). (Excavations should have slopes of no more than 1 foot of rise for 5 feet of run to prevent such concentration along shorelines.) In short, deepening significant areas in small restored wetlands is often counterproductive. Maintaining areas of open water at all times is best done by restoring wetland complexes. In complexes, small restored wetlands provide shallow feeding areas while nearby large semi-permanent and permanent marshes provide open water areas.

Excess fill from excavations can be used to build nesting and loafing sites on deep marshes for ducks and geese. These islands must be made inaccessible to predators to be effective. Unfortunately, this is hard to do and could be impossible in smaller restorations. In any case, islands will always be accessible to avian predators. At Union Slough National Wildlife Refuge in northern Iowa, nesting islands did not increase hatch rates of waterfowl because they are in shallow wetlands accessible to raccoons and mink (Fleskes and Klaas, 1991). Local sportsmen constructed islands surrounded by deep (more than 6 feet) and wide (greater than 8 feet) moats in a large restored marsh in Redwood County, Minnesota. Their design seems to be more effective at stopping predation by mammals. In general, islands should have a minimum diameter of 20 feet, be located at least 50 feet from the shore, and be surrounded by water at least 2 feet deep (Zenner et al., 1992). Newly constructed islands should be seeded with a temporary cover crop.

CASE STUDY ━━━━━━━━━━━━━━━━━━━━━━━━━━━━━━━━━━

WETLAND DESIGN

In 1988, the Iowa Department of Natural Resources acquired a 142-acre tract in Dickinson County, Iowa, that had been enrolled in the Conservation Reserve Program for two years. The property, now called "McBreen Marsh," includes seven historic wetland basins: three semi-permanent basins (Blue Earth series) and four seasonal basins (Okoboji series). Several large lakes surround the elevated, hilly area where McBreen Marsh is situated. The site had been extensively tiled in 1968 and used primarily for corn production. The state decided to restore six basins on the site by interrupting tiles and installing water control structures. The overall restoration design for this project is given in Figure 4.10. Construction was completed during the summer and fall of 1988.

A single 6-inch tile line originated in Basin A and drained into Basin C. Basin A was restored by replacing 50 feet of clay tile with impervious tile. Since the basin is close to a property boundary, a maximum pool elevation needed to be established to ensure nearby cropland would not flood. A nonperforated tile riser was created at the outlet end of the new tile section. The top of the riser is below the elevation of the property line. Water fills the basin until it reaches the top of the riser; then it flows into the tile and onto Basin C.

Basin B is a 7.5 acre semi-permanent basin drained by five parallel tile lines that connect to an 8-inch tile main. Fifty feet of clay tile along the main line was replaced with impervious tile. However, unlike Basin A, Basin B is a deeper basin and could require drawdown management to periodically rejuvenate emergent vegetation. A gate valve water control structure was installed rather than a simple tile riser to allow for

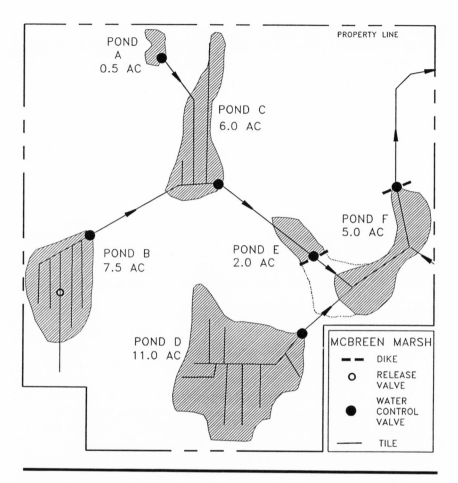

Figure 4.10. Restoration design for the McBreen Marsh complex (T100N R37W S13) in Dickinson County, Iowa.

more precise water management. Gate valves, used in municipal water mains, allow a continuous range of maximum water depths. A standpipe was also installed at the location of an existing surface inlet, near the center of the basin. During cultivation, the surface inlet allowed ponded water to flow into the line without needing to first percolate through the soil. After restoration, the standpipe relieves back pressure in the flooded lines during high flow periods. Back pressure could cause the tile to rupture upstream of the basin if a pressure release valve was not present to allow water to exit to the surface.

Basin C receives flow from Basin A and B and is internally drained by three parallel tile lines. The 8-inch line exiting Basin C flows to Basin E. Basin C is a 6-acre, semi-permanent basin. Fifty feet of clay tile along the main line in Basin C was replaced with impervious tile and an irrigation valve, DOS-IR valve, was installed for water control. DOS-IR valves are lighter and less expensive than gate valves yet still permit a continuous range of maximum water depths. No pressure release valve was necessary on Basin C because the slope of incoming tile lines was gradual enough to make back pressure problems unlikely.

Basin D is the largest of the McBreen Marsh wetlands, 11 acres. At least seven randomly situated tile lines drain into an 8-inch main, which connects to Basin F. Like Basin B, clay tile was removed along the main line and replaced with impervious tile, and a gate valve was installed at the outlet.

Basin E and F are actually part of a single historic wetland basin. A preconstruction survey indicated that the original basin would be a shallow, 8-acre wetland. An evaluation of its full pool elevation showed that much of the area, including the entire western section, would be very shallow (less than 1 foot at full pool). Wildlife biologists decided to increase the depth of the wetland by constructing a dike across the western section and another at the natural outlet. Impervious tile replaced clay tile beneath the dikes for both basins, and irrigation valves (DOS-IR valves) were installed.

The McBreen Marsh restorations illustrate common design features used in the prairie pothole region including tile replacement, tile risers to limit pool elevation, water control structure installation, pressure release valves to avoid upstream tile line ruptures, and dike construction to increase basin depth. McBreen Marsh differs from many other restorations in the region because wildlife biologists who planned the projects generally chose to restore wetlands to recreate the historic conditions of the site. Consequently, dikes were only constructed on two basins, and only one basin (E) was designed to be a different type of wetland than before drainage. Water control structures also were placed on all wetlands where drawdown management may be desirable in the future. The result is a complex of restored wetlands that includes temporary to semi-permanent wetlands.

WATERSHED LAND USE

Soil erosion from surrounding hillsides has resulted in the gradual filling of prairie wetlands with sediment since the melting of the glaciers. Since European settlement and the plowing up of the prairies, soil erosion rates have increased dramatically. Many drained basins that are being restored, especially those in rolling terrain, are today much shallower than they were 100 years ago because they have been partially filled by sediment from surrounding farm fields. Although wetlands are often touted as sediment traps, wetlands are not an ideal solution for soil erosion problems. Natural and restored wetlands are severely degraded by inputs of sediment and associated nutrients and pesticides. The potential life span of natural and restored wetlands will also be significantly reduced in watersheds with severe erosion problems. Farm practices that reduce soil erosion on-field are good for both farms and wetlands. The key to maximizing life spans of restored wetlands is reducing sediment inputs. Site evaluations should include a consideration of potential sediment inputs into the restored basin, i.e., the potential life span of the basin. All else being equal, basins with lower potential inputs should be given higher priority for restoration. Sites with significant soil erosion problems do not have to be eliminated from consideration for restoration, but as part of the restoration project suitable adjustments in the design of the project need to be made and possibly in the farming practices in the basin.

The size of watershed, kind of land cover, and steepness and length of slopes leading to the restored basin are used to calculate sedimentation rates into the basin. See Table 4.2 for soil loss rates and sediment values to be expected for most conditions typically found in the region, and Table 4.3 for how to calculate sedimentation rates. Soil loss rates used in these calculations are based on estimates from the USDA Universal Soil Loss Equation. Local SCS personnel can help with calculating soil loss estimates for specific projects. If sedimentation is likely to have been excessive in the past, excavating some of these sediments during construction may be appropriate (see previous section).

The amount of sediment entering a wetland can be reduced by having a perennial grass buffer around the restored wetland and by implementing appropriate soil conservation practices in the watershed. All restored wetlands should have a buffer of permanent, close-growing vegetation, such as prairie grasses, around the basin to reduce sediment inputs as well as nutrient and pesticide inputs. Permanent grassland (e.g., hay fields, pastures, prairies) is the most effective way to reduce

TABLE 4.2. Selected soil erosion estimates for a range of conditions commonly encountered within the southern prairie pothole region. Soil Conservation Service state office staffs in Iowa, Minnesota, and South Dakota made these calculations from the Universal Soil Loss Equation.

Location	Situation	Soil Loss Rate (Tons/Acre/Year)
CENTRAL MINNESOTA Kandiyohi County	Corn-soybean rotation, fall moldboard plow.	10.5
Typical Soil: Sundburg-Wadenhill Slope: 7-10%	Corn-soybean rotation mulch till with 30% residue after planting corn and 40% residue cover after planting soybeans.	5.3
	Grazed pasture.	0.3
	Unharvested permanent grass pasture.	0.1
SOUTH-CENTRAL MINNESOTA Cottonwood County	Corn-soybean rotation fall mulch till/moldboard plow leaving 10% residue cover after planting corn and 10% residue cover after planting soybeans.	3.7
Typical Soil: Clarion-Nicollet Slope: 2-5%	Corn-soybean rotation mulch till leaving 25% residue cover after planting corn and 30% cover after planting soybeans.	2.6
	Grazed pasture.	0.2
	Unharvested permanent grass cover.	0.04
SOUTHEASTERN SOUTH DAKOTA Brookings County	Corn-soybean rotation fall moldboard plow.	18.2
Typical Soil: Poinsett-Buse-Waubay Slope: 7-10%	Corn-soybean rotation leaving 20% ground cover after planting corn and 35% ground cover after planting soybeans.	9.7
	Grazed pasture.	0.6
	Unharvested permanent grass pasture.	0.1
SOUTHEASTERN SOUTH DAKOTA Brookings County	Corn-soybean rotation, fall moldboard plow.	8.0
Typical Soil: Poinsett-Buse-Waubay Slope: 2-5%	Corn-soybean rotation leaving 20% ground cover after planting corn and 35% ground cover after planting soybeans.	4.3
	Grazed pasture.	0.3
	Unharvested permanent grass pasture.	0.05
NORTHWESTERN IOWA Dickinson County	Corn-soybean rotation, fall moldboard plow.	21.0
Typical Soil: Clarion-Nicollet Slope: 7-10%	Corn-soybean rotation mulch till with 30% residue after planting corn and 40% residue cover after planting soybeans.	15.0
	Grazed pasture.	1.0
	Unharvested permanent grass cover.	0.2
NORTH-CENTRAL IOWA Pocahontas County	Corn-soybean rotation, fall moldboard plow.	7.0
Typical Soil: Clarion-Nicollet Slope: 2-5%	Corn-soybean rotation mulch till with 30% residue after planting corn and 40% residue cover after planting soybeans.	5.0
	Grazed pasture.	0.3
	Unharvested permanent grass cover.	0.1

TABLE 4.3.   Sample calculations for estimating project life span. A 2-acre re-stored basin in Brookings County, South Dakota, is within an 18-acre watershed. Approximately 6.5 acres of the watershed are in corn-soybean rotation, leaving 20%–35% residue; 9.5 acres are in permanent cover (Conservation Reserve Program). The slope gradient of the crop field was 5% and the slope gradient of the CRP area was 8%.

Crop Area: 6.5-acre	CRP Area: 9.5-acre
Soil Loss Rate (from Table 5.1): 4.3 tons/acre/yr	Soil Loss Rate: 0.1 tons/acre/yr
Sediment Volume (loss rate x 21.73): 93.44 cu ft/acre/yr	Sediment Volume: 2.17 cu ft/acre/yr

Sediments from Crop Area per year = 6.5 acre x 93.44 cu ft/acre = 607.4 cu ft
Sediments from CRP Area per year = 9.5 acre x   2.17 cu ft/acre =   20.6 cu ft

Total Sediments from watershed = 607.4 + 20.6 = 628 cu ft/year
Volume of Basin (from topographic survey): 44612 cu ft.

PROJECTED LIFE SPAN OF BASIN (if current land use persists):
44612 cu ft (volume of basin) / 628 cu ft/ year = 71 YR

soil erosion because a tight grassland sod will has 95% less soil erosion than a cropfield (U.S. Soil Conservation Service, 1983). If the grassland is grazed, soil erosion will be greater. Proper grazing management is essential for maintaining adequate vegetative cover. Grassland buffers are also important for many species of wildlife such as dabbling ducks who make their nests in uplands. The wider this buffer, the better. The minimum buffer width that should be considered is 100 feet.

Particularly in watersheds with steeps slopes and in row crops efforts should be made as part of the wetland restoration plan to introduce soil conservation practices in the watershed. These practices include contour farming and terracing. Contour farming involves planting crops perpendicular to the direction of the slope, whereas terracing involves creating step-like gradations along the slope face. In both cases, soil is trapped on the field and not permitted to travel far from its point of detachment. Contouring is most effective on slopes of 2%–7%, reducing erosion 40%–50% (U.S. Soil Conservation Service, 1983). When combined with conservation tillage, soil erosion can be reduced by 70%–80%. Contour farming is much less effective on steeper slopes. Terraces are appropriate in these situations and will have an equally dramatic effect on limiting soil erosion by 60%–90%. Changes in tilling practices should also be encouraged. Conventional tillage results in practically bare soil much of the spring and fall, generally the periods of highest rainfall in the region. In contrast, conservation tillage, including chisel planting, strip tillage, till planting, plow planting, or wheel track planting, leaves crop residues on the field. This protective cover minimizes soil erosion by allowing more water to infiltrate rather than run off and reducing raindrop impact soil detachment.

REFERENCES

Anderson, G.R. 1985. Design and location of water impoundment structures. IN M.D. Knighton (Ed.). Water Impoundments for Wildlife: A Habitat Management Workshop. U.S. Forest Service General Technical Report NC-100.

Atlantic Flyway Waterfowl Council. 1959. An illustrated small marsh construction manual based on standard designs. Vermont Fish and Game Service. Montpelier, Vermont.

Bishop, R.A., R.D. Andrews, and R.J. Bridges. 1979. Marsh management and its relationship to vegetation, waterfowl and muskrats. Proceedings of the Iowa Academy of Science 86: 50–56.

Brooks, K.N., P.F. Folliott, H.M. Gregersen, and J.L. Thames. 1991. Watershed considerations for engineering applications. IN *Hydrology and the Management of Watersheds*. Iowa State University Press, Ames.

Fleskes, J.P. and E.E. Klaas. 1991. Dabbling duck recruitment in relation to habitat and predators at Union Slough National Wildlife Refuge, Iowa. U.S. Fish and Wildlife Service, Fisheries and Wildlife Technical Report 32. 19 pp.

Fredrickson, L.H. 1985. Managed wetland habitats for wildlife: why are they important? *IN Water Impoundments for Wildlife: A Habitat Management Workshop*. North Central Forest Experiment Station, U.S. Forest Service, St. Paul, Minnesota.

Fredrickson, L.H. and F.A. Reid. 1986. Wetland and riparian habitats: a nongame management overview. IN J.B. Hale, L.B. Best, and R.L. Clawson (Eds.). *Management of Nongame Wildlife in the Midwest: A Developing Art*. Proceedings of a Symposium of the 47th Midwest Fish and Wildlife Conference. Grand Rapids, Michigan.

Olson, R. 1990. Developing artificial wetlands to benefit wildlife and livestock. University of Wyoming Extension Publication B-938.

Stewart, R.E. and H.A. Kantrud. 1971. Classification of natural ponds and lakes in the glaciated prairie region. Research Publication 92. Bureau of Sport Fisheries and Wildlife.

U.S. Soil Conservation Service. 1962. Iowa drainage guide. Published with Iowa State University as Agriculture and Home Economics Experiment Station Special Report 13.

U.S. Soil Conservation Service. 1975. Guidelines for soil and water conservation in urbanizing areas of Massachusetts. U.S. Department of Agriculture, Washington, D.C.

U.S. Soil Conservation Service. 1983. Water quality field guide. Technical Publication No. 160. U.S. Department of Agriculture. Washington, D.C.

U.S. Soil Conservation Service. 1991. Wetland restoration, enhancement, or creation. IN *Engineering Field Manual*. Washington D.C.

Wenzel, T. 1992. *Minnesota Wetland Restoration Guide.* Minnesota Board of
    Water and Soil Resources. St. Paul, Minnesota.
Zenner, G.G., T.G. LaGrange, and A.W. Hancock. 1992. Nest structures for
    ducks and geese: a guide to the construction, placement, and maintenance
    of nest structures for Canada geese, mallards, and wood ducks. Iowa De-
    partment of Natural Resources, Des Moines. 34 pp.

EVALUATION AND MANAGEMENT

What criteria should be used to determine if a restored wetland project has been successful or not? This is an important question that seems to be rarely asked. It is important for two major reasons: (1) it is only by evaluating restoration projects that information needed to improve project design and management can be obtained, and (2) a set of criteria is the only way to determine if restored wetlands are similar to and functioning like natural wetlands as the proponents of wetland restoration claim. How closely restored wetlands resemble and function like natural wetlands is the definitive test of their success. There are major obstacles to be overcome when evaluating wetland restorations. One, there is often inadequate information about the composition and functioning of natural wetlands. Two, each class of restored wetlands should be compared to only the same class of natural wetlands, and determining the class of a restored wetland can be difficult. Three, some natural wetlands change cyclically in composition and function, making evaluations more complex. Four, functional assessments can be very expensive and require years to complete. Five, it must be determined when such an assessment should be made: restored wetlands change rapidly.

There is no fixed number of years after a restoration that an assessment should be made. Restored wetlands are in an early stage of succession during which a new hydrology is being established and plant and animal species are recolonizing the sites. These changes will be most rapid and dramatic in the first few years and then should begin to slow down. The rate of change is also affected by weather conditions. During drought years, it is especially slow. Two or more years with normal or above-normal annual precipitation are needed before the water regime can be determined and before a significant amount of revegetation will occur. The best way to evaluate a restored wetland is by making simple, routine annual evaluations until the wetland's features stabilize. We recommend that, as a minimum, information be collected on water depth and vegetation in all restored wetlands for at least five years and then every two or three years. Even the most casually collected and superficial information about a restored wetland is better than none. The most important reason to do annual evaluations is not to determine if a site has been successfully restored or not but to determine if any unforeseen problems have developed. It is only by going back to a restored wetland and recording what has happened that any problems with the restoration can be discovered and corrected. Often the earlier problems are discovered, e.g., a leaking dike, the more easily they can be corrected.

There are only a few features of natural wetlands and restored wetlands that can be routinely used to evaluate their success: water regime, composition of the vegetation, and some aspects of wildlife use. We cannot assess the functioning of restored wetlands easily for water quality improvement, flood storage, or most aspects of wildlife habitat. Because of limitations in the amount of data that can be collected, routine wetland evaluations in the foreseeable future will be done primarily by comparing lists of species in restored wetlands with lists of species in natural wetlands. For the time being, we will have to assume that the more a restored wetland resembles a natural wetland in features, the more it also resembles it in function. Of the three features that can be used to evaluate the success of a restoration, the two simplest are also the most important, water regime and vegetation. These two features determine to a large extent how a restored wetland will function as wildlife habitat, for improving water quality, and for flood storage. This chapter is divided into two sections. The first (evaluation) outlines some simple procedures for collecting information on restored wetlands and for determining how closely a restored wetland resembles comparable natural wetlands. Follow-up visits to restored wetlands will also provide invaluable information about when and how restored wetlands revegetate and are recolonized by different groups of animals. At the moment, very little is known about the development of restored wetlands. The postrestoration data that should be taken routinely are generally easy, simple, and quick to collect. Some data collection on wildlife use, however, does require fairly elaborate censusing techniques. The second section (management) deals with the kinds of problems that have been detected as a result of evaluations and suggests possible solutions for them.

EVALUATION
Water Regime

Determining the water regime on a site can be as simple as visiting the site every few months and noting if standing water is present. This should be done for five years or until the nature of the normal water regime is established. If more detailed information is needed, simple water staff gauges can be made of sheet metal and t-posts with brightly colored depth marks painted on them so that they can be read at a distance. Commercial versions of these gauges are also available. These gauges should be placed in a deep part of the basin. Unfortunately, gauges often get obscured after a few years by stands of tall vegetation or may become unreadable if algae coat the gauge surface. Eventually, water depth readings often will require a trip into the wetland. Permanent records of these water levels must be kept in order to help diagnose any problems that may develop with the restorations.

The water regime of a restored wetland, once it is known, can be used to place the restored wetland in the correct Stewart and Kantrud pothole class (see Chapter 2). Assigning the restoration to the correct class is the first step in the evaluation process. Whether the restored wetland does or does not have the hydrology that was expected on the basis of the site evaluation should be used as a minimum criterion to determine if a restoration should be classified as a success, partial success, or failure. If the restored wetland has the hydrology, i.e., is the type of wetland expected, it should be considered a success. If the restored wetland is in an adjacent class, i.e, a semi-permanent wetland was expected but a seasonal wetland developed, it should be classified as a partial success. Restorations whose hydrology do not resemble that expected should be classified as failures. If nothing else, restored wetlands that never hold any water should be classified as failures and not included in counts of restored wetlands in the region, something that is currently not done.

Water Regime

A prominent terminal moraine straddles the southern edge of Winnebago County, Iowa. The wetlands to the north of this moraine historically included large deep lakes with expanses of wild rice along their periphery, shallow lakes, sedge meadows, and smaller prairie potholes. Restorations in this county have been initiated to restore all of these wetland types with varying success.

One of the basins restored is a 2-acre wetland with an Okoboji (muck) series soil, indicating the wetland was historically a semi-permanent wetland. This wetland was tile drained approximately 50 years before restoration. The landowner occasionally observed spring flow at one location within the basin after tile drainage. The surface watershed is estimated at 37 acres from topographic and soil survey maps. The ratio of watershed size to basin size (see Figure 3.10) suggests that this basin should hold water in most years. The basin flooded shortly after construction and has retained water since. The wetland held 13% of its full pool volume into midsummer during the drought year of 1989 and was 70% to 80% full in July and August of 1990 and 1991, wetter years. This restoration should be considered a success.

Another basin restored is a 14-acre wetland that has a Houghton series soil. The soil indicates that this former wetland was typically flooded. Over 3 feet of organic deposits with visible plant fragments were observed in a soil core taken from this basin, confirming the soil survey map information. Yet the basin has not retained water since construction. At best, a few inches of standing water occur briefly in some areas after a heavy rain. The restored wetland is actually an arm of a much larger drained shallow lake, formerly called Bear Lake. Its very sinuous basin extended over approximately 800 acres, and its outflow entered Bear Creek, a tributary of the Winnebago River. According to area residents, the former lake was 4 to 5 feet deep in places and was a spawning ground for northern pike that traveled to the lake through Bear Creek from the Winnebago River. Residents report the lake was drained in the late 1920s and early 1930s, first by a series of county ditches and later by tile. Some tile drains were even outfitted with motorized pumps. The restored part of the basin was first tiled in 1972 and then retiled in 1975. The surface watershed of the entire drained lake is small, and the estimated watershed for the restored portion of the basin is only 32 acres. The watershed to basin area ratio suggests this basin should often retain water. It may be that this lake historically received most of its water from groundwater discharge, rather than from surface runoff. If the lack of groundwater flow is the reason this restoration does not have surface water, this restoration may never have standing water except briefly after major storms or after snowmelt because of the lowering of the regional groundwater level. An additional complicating factor could be that most of Bear Lake is still drained. It is unclear how much lateral seepage occurs through the peat from the restored wetland to the remaining drained portions of the lake. Restoring only part of a larger drained basin will require care to ensure that the restored portion will have a hydrology compared to that in its predrained condition. Unless this basin does eventually reflood, this restoration will have to be classified as a failure. If sufficient numbers of restorations are done in the area and the groundwater table returns to its presettlement levels, this project might eventually become a success.

Vegetation Composition

The species composition and structure of the vegetation of restored wetlands is generally their easiest feature to document. The species composition of restored wetlands can also be easily compared to that of natural wetlands to assess their similarity. Detailed studies of the vegetation of restored wetlands are not needed. All that is needed is a list of species. The vegetation of recently restored wetlands is typically not well organized into zones, which should develop with time. Consequently, lists of species present in the entire basin are the best means to assess the status of the vegetation. Additional information, however, is often useful for interpreting the status of the vegetation, especially photographs of the basin or a section of the basin taken from the same vantage point and at the same time each year. Instead of photographs, simple, hand-drawn vegetation maps of the basin can be made.

A fairly complete plant list is required to evaluate the vegetation of a restored wetland. This list can be compiled by thoroughly searching the basin three times during the year—in midspring (May), midsummer (July), and late summer (September). Three visits will ensure that all species will be seen and can be identified. Only one or two visits may be needed if the observer knows the local wetland flora. A plant list can be completed in a few hours or less if the plants are familiar to the observer.

Appendix A1 lists all species known from the region and their designated group. Rare species that could be found in restored wetlands are listed by county in Appendix A2. Keys for identifying common wetland species in the southern prairie pothole region are found in Appendix A3. A data sheet for recording species lists is given in Appendix D8. The cover of each species should also be noted, if possible. A standard cover-abundance scale used by plant ecologists is given in Appendix D9. Changes in the number of species and their cover can be used to monitor the development of vegetation in a restored wetland, particularly if field data were collected at the same time of year and with a similar amount of effort each year. Monitoring every two to three years after the first few years should be adequate.

Species of plants in freshwater prairie potholes belong to 10 groups on the basis of their growth form, life span, and water depth or flooding tolerance (Table 5.1). Plants within a group respond somewhat similarly to hydrological conditions within a wetland. Some groups, such as submersed aquatics (Group SA), woody plants (Group WO), and floating annuals (Group FA), are characterized by their distinctive growth form whereas others by their life span—mudflat annuals (Group MA). Many marsh plants are herbaceous, upright perennials. These can be classified by their tolerance to flooding. Species that cannot tolerate flooding into the growing season are in the wet prairie (WP) and sedge meadow (SM) groups. Species in shallow and deep emergent perennials, groups SE and DE, can withstand shallow flooding into the growing season (SE) or perpetually (DE). These groups do not correspond to the vegetation zones encountered in prairie wetlands. Species within a zone often belong to several groups, but species from one group typically dominate a zone. For example, a sedge meadow zone in a natural wetland may contain primarily Group SM plants but also normally has many Group SE plants. The vegetation of restored wetlands is best evaluated by examining the number of species present in each group. Not all groups, however, will be found in every wetland. Ephemeral and temporary wetlands (Class I and II) will likely only have species in groups WP, SM, and perhaps MA and WP whereas semi-permanent and permanent marshes (Class III and IV) will also have SE, DE, SA, and FA species.

To evaluate the vegetation of a restored basin, the number of species found in each group should be compared to that found in the same groups in natural wetlands of approximately the same size. Detailed vegetation data from 10 natural wetlands as well as published plant lists from other natural wetlands in the region were used to establish the number of species expected for each group in natural wetlands (Galatowitsch, 1993). The number expected in depauperate, typical, and excep-

TABLE 5.1. Groups of common species found in prairie potholes. Different groups are characterized by similar growth form, life span, and environmental tolerance. Scientific names are given in Appendix A.

Group	Characteristics	Common Plants
WP	Perennial plants that cannot tolerate more than a few weeks of flooding in the spring and so are confined to wet prairie areas. Group WP is largely grasses and forbs.	Big bluestem, switchgrass, Kentucky bluegrass, cow parsnip, golden alexander, field thistle, Flodman's thistle, sunflower, blazing star, prairie gayfeather, gray coneflower, compass plant, cup plant, goldenrod, fleabane, spiked lobelia, tick clover, mountain mint, wild garlic, yellow star grass, prairie phlox, Canada anemone, tall meadow rue, culver's root.
SM	Perennial plants that do not occur in areas flooded more than one to two months in the spring. Most of these plants are most abundant in sedge meadows but are often common in wet prairies. Group SM species include sedges, rushes and some forbs.	Woolly sedge, awn-fruited sedge, tussock sedge, foxsedge, nutgrass, spike rush, green bulrush, field horsetail, Dudley rush, Torrey's rush, bluejoint, rice cut grass, cord grass, fowl meadow grass, water hemlock, indian hemp, swamp milkweed, asters, beggar ticks, joe pye weed, boneset, sneezeweed, sawtooth sunflower, ironweed, water horehound, northern bugleweed, field mint, common skullcap, woundwort, curly dock, prairie loosestrife, marsh marigold, monkey flower, water speedwell, blue vervain.
SE	Perennial plants that can tolerate shallow flooding extending into mid-summer during most years. These plants, typically water plaintains, arrowheads, leafy bulrushes, and large sedges, are usually most abundant in shallow emergent zones.	Water plantain, wapato, awned-sedge, lake sedge, beaked sedge, large spike rushes, river bulrush, American sloughgrass, tall manna grass, fowl manna grass, giant burreed, broad leaf cattail, water parsnip, blue flag, water smartweed, tufted loosestrife.
DE	Perennial plants that can withstand shallow flooding for several years. These are tall emergent plants such as some bulrushes that occur at the edges of open water in semi-permanent marshes.	Hard-stem bulrush, soft-stemmed bulrush, reed, narrowleaf cattail, hybrid cattail.
SA	Submersed aquatic plants such as pondweeds and bladderwort that grow within the water column. Some species are only found in open water whereas others, such as bladderwort, can grow in the understory of emergent aquatics.	Coontail, common bladderwort, water milfoil, bushy pondweed, leafy pondweed, long-lived pondweed, sago pondweed, baby pondweed, flat-stemmed pondweed.
FA	Floating annual plants, like duckweeds, that are small free floating plants occurring at or immediately below the water's surface.	Lesser duckweed, greater duckweed, watermeal, star duckweed, slender riccia, purple-fringed riccia.
MA	Mudflat annuals common to muddy unvegetated shorelines in restored marshes and drawdown areas in the center of semi-permanent marshes. Many agricultural weeds are in Group MA.	Yellow nutgrass, fragrant cyperus, annual spike rushes, ticklegrass, redtop, barnyard grass, teal lovegrass, foxtail barley, common witchgrass, pigweed, sticktight, beggar ticks, spiny cocklebur, marsh cress, lambs-quarters, nodding smartweed, pinkweed, golden dock, cursed crowfoot.
WO	Woody plants, such as willows.	Cottonwood, peach-leaved willow, sandbar willow.
FN	Perennial plants that grow in fens - calcareous meadows that receive groundwater flow and so are nearly always saturated but not flooded. Many of these species also occur on sedge meadows (Group SM), but some are restricted to very mineralized sites.	Water sedge, inland sedge, tussock sedge, shining cyperus, spike rush, tall cotton grass, large arrow grass, bog rush, northern reed grass, muhly, Riddell's goldenrod, cinnamon willow herb, bog violet, bog bean.
FP	Floating rooted perennials such as water lilies which may have been more common in the past in large semi-permanent marshes and lakes but are not often encountered.	Yellow water lily, white water lily.

tional natural wetlands for selected groups is given in Table 5.2. By comparing the number of species in a group in a restored wetland with those in different quality natural wetlands in Table 5.2, the status of the group in the restored wetland can be determined. The vegetation of restored wetlands ranging in size from 1 to 20 acres can be evaluated using Table 5.2. No comparable tables are currently available for either smaller or larger natural wetlands.

TABLE 5.2. The number of species in different groups of wetland plants in depauperate or poor, typical and exceptionally good natural potholes in the southern prairie pothole region.

Species Group	Total Number of Species		
	Depauperate	Typical	Exceptional
WP	1-5	6-14	15 +
SM	1-15	16-30	31 +
SE	1-6	7-15	16 +
DE	1-2	3-4	5 +
SA	1-3	4-10	11 +
FA	1-2	3-5	6

Source: Adapted from Galatowitsch, 1993.
Note: Species groups are defined in Table 5.1.

For each group in Table 5.2, the restored wetland should be given 0 points if no species were present in a group, 1 point if the number of species places it in the depauperate category, 2 if it is in the typical category, and 3 if it is in the exceptional category. Thus, the vegetation of restored Class I and II wetlands could have an overall rating of 0, no vegetation, to 6, comparable to the best natural example of these classes. Only groups WP and SM should be considered when evaluating the vegetation in restored wetlands in these classes. If the vegetation rating is <2, the vegetation of the restored wetland should be classified as poor. If it is 3 or 4, it should be classified as good. If it is >4, it should be classified as excellent. The vegetation of a restored Class III or Class IV wetland could have an overall rating from 0, no vegetation, to 18, comparable to the best natural example of these classes. If the vegetation rating is <6, the revegetation of the restored wetland should be classified as poor. If it falls between 6 and 8, it should be classified as fair, and between 9 and 11 as good. If it falls between 12 and 18, it should be classified as excellent.

Problems with revegetation could be serious enough that appropriate management action should be taken if one of the following occurs:

1. Little or no revegetation after three or more years, i.e., a vegetation rating of < 6 for a Class III or IV wetland or < 2 for a Class I or II.
2. One or more vegetation types normally found in a wetland of that class fails to develop.
3. Aggressive weeds listed are present.
4. Woody species, i.e., species in group WO, are present.

Management recommendations indicated by situations are outlined in the management section of this chapter. Restored wetlands will be of particular ecological significance and may require special management also if the following species occur:

1. Any rare plant species are found (see Appendix A2).
2. At least five fen species (Group FN) are found.

It should be noted that the evaluation scheme recommended is a minimal evaluation since no attention is paid to the abundance of species present in the restored wetland or to the structure of the vegetation, e.g., the presence or absence of well-defined zones. Many more sophisticated techniques are available for comparing the composition of restored and natural wetlands in the plant ecological literature. The field data and statistical analyses required for these are normally beyond the scope of routine site evaluations. For situations where more sophisticated evaluations of the development of vegetation are warranted, standard texts on vegetation sampling and analyses such as Ludwig and Reynolds, 1988, or Causton, 1988, should be consulted.

CASE STUDY ━━━━━━━━━━━━━━━━━━━━━━━━━━━━━━━━━━━━━━━

Revegetation

The effects of site history on revegetation can be seen by comparing the revegetation of two basins of similar size and depth restored in 1988. One basin is a 5-acre wetland in Palo Alto County, Iowa, with a mean depth of 1.76 feet. The other is a 4-acre wetland in Kandiyohi County, Minnesota, with a mean depth of 1.62 feet. Both sites were in row crops before restoration and had been drained by tile. Both sites flooded immediately after tiles were broken and have been flooded since, except for a brief period in the late summer of 1989, when both sites were only saturated in the central portion of their basins. Their drainage histories, however, are quite different. Drainage tile was installed in the Iowa site in 1937, and it was planted in row crops. The Minnesota site was tiled in 1983 and cultivated for only five years. The Minnesota basin should have had the richer seed bank at the time of restoration (Wienhold and van der Valk, 1989). Also, some rhizomes of wetland species may have survived this brief period of drainage and cultivation in the Minnesota wetland.

After one year of restoration, 8 wetland plants had colonized the Iowa basin (Table 5.3). By 1991, three years after ponding, 22 species

TABLE. 5.3. Cover of plant species in a wetland restoration in 1989-1991 in Palo Alto County, Iowa. (Cover values are described in Appendix D. Invasive or weedy species are indicated with a √.)

SPECIES		COVER VALUES		
		1989	1990	1991
Wet Prairie Species (WP)				
Quack Grass	√	1	1	2
Sedge Meadow Species (SM)				
Canada Thistle	√	R	2	R
Shallow Emergent Species (SE)				
Water Smartweed		R	R	R
Wapato			R	
Water Plantain				R
River Bulrush				R
Reed Canary Grass	√			R
Deep Emergent Species (DE)				
Cattails	√	R	2	3
Soft-stemmed Bulrush				R
Submersed Aquatics (SA)				
Leafy Pondweed			3	3
Sago Pondweed			1	R
American Pondweed			R	R
Common Bladderwort			3	3
Baby Pondweed			2	R
Coontail				R
Bushy Pondweed				1
Flat-stemmed Pondweed				R
Floating Annuals (FA)				
Lesser Duckweed		R	R	1
Greater Duckweed		R		1
Star Duckweed			R	R
Mudflat Annuals (MA)				
Pennsylvania Smartweed		1	1	
Curly Dock		2	R	R
Nodding Smartweed				R
Beggars-tick				3

were present. Only invasive weeds, quackgrass and Canada thistle, were established in areas that would potentially be wet prairie and sedge meadow. Five shallow marsh species were present in 1991, but none had attained a significant cover. Hybrid and narrow leaf cattail were established on the site in 1989 and expanded to cover approximately three-quarters of the flooded portion of the basin by 1991. A small patch of soft-stemmed bulrush was established in 1991. Submersed aquatics were established by the second year of flooding. Eight species, including two abundant pondweeds, were present on the site by 1991. Duckweeds were present on the site soon after restorations and were present in low abundance during each year monitored. Two to 3 mudflat annuals, commonly present in crop fields, were present each year as well. The vegetation rating of this wetland was 6, 7, and 8 in 1989, 1990, and 1991, respectively, i.e., fair in all three years.

The Minnesota basin was considerably more diverse (Table 5.4). Nineteen species were present on the site in 1989; by the third year, 36 species had colonized. Still, wet prairie plants were only represented by the weedy quackgrass. Twelve sedge meadow species occurred on the site, most establishing during the second and third year after reflooding. Shallow emergent species gradually established on the site, with 4 species present in 1989, and 7 species present in 1990 and 1991. The same 2 deep water marsh species are present as in the Iowa site, but soft-stemmed bulrush, which established within the first year after restoration, is more common than cattail. Like the Iowa site, submersed aquatics were diverse (6 species) by the third year. However, two species were present in the Minnesota wetland a year after restoration. Duckweeds were present each year in low abundance. Six to 7 mudflat annual species were present each of the first three years. The vegetation rating of this wetland was 5, 7, and 8, respectively, in 1989, 1990, and 1991; i.e., it went from poor in 1989 to fair in 1990 and 1991.

Both restored wetlands have similar vegetation ratings: both are rated fair in 1990 and 1991, and neither resembles a typical natural wetland very closely. However, after three years, the number of submersed (SA) and floating annuals (FA) in the Iowa basin were in the range found in typical natural wetlands. The number of shallow emergent (SE) and submersed aquatic (SA) species in the Minnesota wetland were in the range found in typical natural wetlands. Although their vegetation ratings were similar, these two wetlands have very different species in them, and their most common species are dissimilar.

Except for wet prairie species and floating plants (duckweeds), the Minnesota basin was more rapidly colonized by a more diverse array of wetland plants. These differences were apparent in each of the first

TABLE 5.4. Cover of plant species in a wetland restoration in Kandiyohi County, Minnesota. (Cover values are described in Appendix D. Invasive or weedy species are indicated with a √.)

SPECIES		COVER VALUES 1989	1990	1991
Wet Prairie Species (WP)				
Quackgrass	√	1	2	3
Sedge Meadow Species (SM)				
Swamp milkweed		R		
Willow-leaved dock		R	R	R
Canada thistle	√	2	R	R
Nutgrass (*C. strigosus*)				R
Monkey flower				R
Field mint				R
American bugleweed				R
Rice cut grass				R
Green bulrush				2
Sedge (*C. scoparia*)				1
Sedge (*C. vulpinoidea*)				2
Spike rush (*E. erythropoda*)				R
Shallow Emergent Species (SE)				
Water smartweed		R	2	R
Water plantain		2	3	2
River bulrush		2	2	2
Giant burreed		R	1	2
American sloughgrass			R	R
Spike rush (*E. macrostachya*)			R	1
Tall manna grass			R	
Wapato				R
Deep Emergent Species (DE)				
Soft-stemmed bulrush		1	1	2
Cattail	√	R	R	R
Submersed Aquatics (SA)				
Sago pondweed		R	R	2
Leafy pondweed			R	2
Coontail				2
Common bladderwort				2
Flat-stemmed pondweed				2
Sheathed pondweed				R
Floating Annuals (FA)				
Lesser duckweed		R	1	2
Star duckweed				2
Mudflat Annuals (MA)				
Lambsquarters		R		
Water cress		R	R	R
Nodding smartweed		R	2	R
Cursed crowfoot		R	R	2
Spike rush (*E .obtusa*)		R	R	R
Spike rush (*E. acicularis*)		R		R
Pigweed		R		
Barnyard grass			2	R
Cocklebur			R	
Beggar's tick				R

three years after restoration. Sedge meadow and shallow emergent species showed the most pronounced differences. And, although the same two deep water emergent species were present on both sites, the Iowa wetland was initially colonized by a wind-dispersed plant (cattail), whereas the Minnesota site was colonized by an animal-dispersed plant (bulrush). The bulrush undoubtedly was recruited from the seed bank in the Minnesota wetland because it appeared during the first year of the restoration.

Wildlife Use

Increasing the abundance of wildlife populations, particularly waterfowl, has been the primary motivation behind wetland restorations in the prairie pothole region. While it seems intuitively obvious that restored wetlands will be used by wetland animals, very few restorations have been evaluated to see how much they are being used and by what species of animals. Many landowners report breeding pairs of mallards and blue-winged teal and nesting Canada geese, but it is not known if broods were successfully reared. While it is unrealistic to expect animals that depend on large tracts of unbroken prairie to have become reestablished, even animals with more modest landscape requirements will only become reestablished if there is an adequate food supply and tolerable predation pressure.

Evaluating a restored wetland as wildlife habitat is not easy. There are many problems associated with collecting the required data and interpreting them that make such evaluations much more difficult than hydrological or vegetation evaluations. The more limited a wildlife evaluation is, the more likely it will be meaningful. Evaluations that are restricted to a small number of target species, e.g., dabbling ducks, are easier to carry out, and their results are more likely to be reliable.

Evaluating the use of restored wetlands by various groups of animals requires conducting suitable censuses of the restored wetland to determine which of these animals actually uses the site. A reasonably complete inventory will require a number of visits to the site during the year. All information on wildlife use should be recorded in a log. This information is essential for planning future wildlife management activities (see management section). A suggested wildlife data sheet is given in Appendix D10. The techniques outlined here are standard techniques developed by wildlife biologists that collectively should make it possible to compile a complete list of species using a wetland. Wildlife observations make ideal projects for youth groups such as 4H, FFA, scouts, and local conservation groups, such as bird clubs. Most are easy to do, requiring a minimal amount of special equipment.

Birds are perhaps the most conspicuous animals in a wetland and the easiest to survey. The wetland should be visited at sunrise five or six times from May through early July to survey breeding birds (Figure 5.1). Walk through the wetland, stopping at several locations for 10 minutes, and record all birds seen (binoculars are necessary) or heard. Select observation locations to represent the different kinds of vegetation present on the site. Secretive birds, such as the Virginia rail, sora, least bittern, and American bittern, are not easily detected. Playing continuous tape recordings of calls of secretive birds for several minutes will often elicit their calls. Commercial recordings of bird songs can be used to make continuous tapes of songs of secretive birds on a blank cassette tape.

Nest searches are used to determine if breeding birds are found in a restored wetland. These are done by walking through the wet prairie, sedge meadow, and emergent marsh zones in a zigzag pattern, watching nests and birds flushed from nests. Nest dragging is used to survey grasslands for nests of upland-nesting waterfowl. Two people walk or drive around the periphery of a marsh holding a 100–150-foot plastic line or cable between them (Klett et al., 1986). A third observer records all birds flushed as the line passes over the nest. Some survey crews use all-terrain vehicles to nest drag. The type of nest and number of eggs and nestlings should be recorded. This can be done on the same data sheet used for other bird observations. Nest and egg field guides (Harrison, 1978) are helpful aids.

Reptiles and amphibians are less conspicuous—although some can easily be surveyed. Male frogs and toads chorus to attract mates. Frogs and toads can be identified by their distinctive calls. Tapes are available (see Appendix B) that can familiarize new observers with these calls. Visit the site during breeding periods (Figure 5.2), and record frogs and toads seen and heard. The Iowa Department of Natural Resources has encouraged interested individuals to survey for frogs and toads and contribute their findings to a statewide wildlife survey. Other reptiles and amphibians, including salamanders, turtles, and snakes, are much more difficult to survey. Frequent visits to the site will improve your chances of seeing these animals. Warm, sunny days are particularly good times to find turtles and snakes sunning; warm, windless evenings (dusk) are best for surveying frogs.

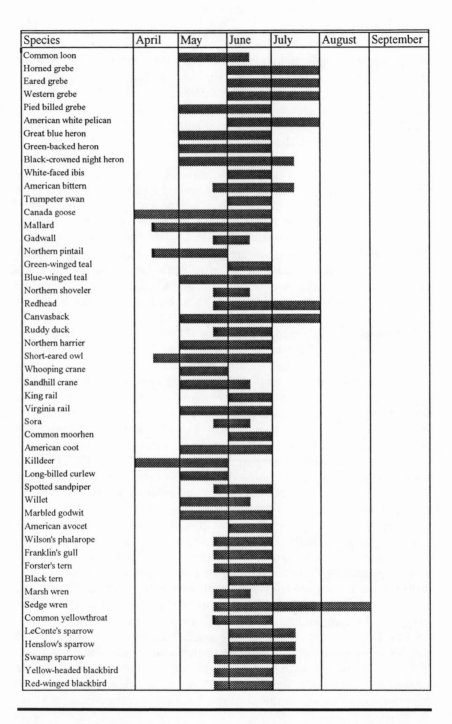

Figure 5.1. Nesting periods for birds in southern prairie pothole region. (From Roberts, 1932)

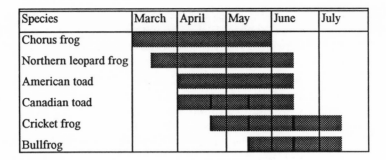

Species	March	April	May	June	July
Chorus frog					
Northern leopard frog					
American toad					
Canadian toad					
Cricket frog					
Bullfrog					

Figure 5.2. Periods when frogs and toads chorus in Iowa. Dates will be somewhat later for central Minnesota and eastern South Dakota. (Data from Christiansen and Bailey, 1991.)

Mammals, like reptiles and amphibians, can be easy or difficult to observe, depending on the species. Some fur bearers, especially muskrats, leave many signs of their activities. Muskrats rip out emergent vegetation, such as bulrush and cattails, and eat the bases, leaving the rest of the plant floating in the water. More noticeable are the lodges they build from these emergent plants or their burrows in dikes. Muskrats can often be seen swimming. Mammals, such as mice, voles, and shrews, are small and secretive. These animals are often sensitive to site quality. Small mammals can be surveyed by setting out a line of small live traps, usually placing two traps every 30 feet, through the sedge meadow and wet prairie periphery of the basin. Traps can be baited with a mixture of peanut butter and oatmeal, and a cotton ball should be placed in each trap to keep captured animals warm. Set traps out during the day, but keep them closed until 6 or 7 P.M. Check traps at midnight and then again between 5 and 6 A.M. Often, several nights of trapping will be necessary to do an adequate survey. Surveying for small mammals requires a great deal of time and effort and cannot be done routinely as part of other wildlife surveys. Such surveys would make good projects for youth or educational groups when supervised by a wildlife biologist.

Because wildlife use of wetlands varies seasonally and from year to year and may be controlled by factors other than wetland or even watershed habitat conditions, the evaluation of a restored wetland as wildlife habitat is more difficult and problematic than are evaluations based on hydrology or vegetation. For restorations done to attract a specific group of animals, the success or failure of a restoration is most easily determined. If the target species use the restored wetland, it can tentatively be classified as a success. If they do not, there could be a problem. This problem could range from the restored wetland never developing the hydrology expected, to the vegetation zone required for the target species never developing, to land use in the watershed not being suitable. In other words, water regime and vegetation information about the restored wetland are also needed to interpret the wildlife data and to determine if remedial action can be taken. If habitat conditions, in fact, seem suitable for the target species, it is also possible that these species have simply not yet colonized the restored wetland or are using the wetland too briefly to be detected. The presence of target species, however, is not sufficient evidence to conclude that a restored wetland is actually good wildlife habitat. More sophisticated studies would need to done on food availability, predation rates, and breeding success to be absolutely sure that the restored wetland actually benefits the populations of target species. It is conceivable that restored wetlands may in fact be detrimental for populations of some species because animals are attracted to them but are quickly killed by predators or are unable to raise broods because of inadequate food supplies.

A more general evaluation of a restored wetland as wildlife habitat can be done by comparing a list of animal species that were found in the restored wetland to the list of animal species commonly present in natural wetlands of the same class. The more similar the list of species in the restored wetland to the list of species in natural wetlands, the more successful the restoration. As with evaluations based on target species, the possibility that some species have not yet reached the site should be considered. Even this more general evaluation should only be done for specific kinds of animals, e.g., birds or amphibians. A realistic list of animal species that should be present in natural wetlands of a particular class is not easy to construct because it depends on a variety of factors including the size of the wetland, proximity to other wetlands, surrounding land use, etc. Lists of species that are found in natural wetlands are best developed from surveys of natural wetlands in the area. If this is not feasible, the information presented on animals associated with different kinds of vegetation types or zones (see chapters 2 and 3) can be used to construct a preliminary list of species for a wetland class by compiling a cumulative list of species that should be found in all the zones that characterize a given class of natural wetlands. Preliminary lists should always be verified by experts on the group of animals being examined.

Animal use of restored wetlands can be rare, occasional, seasonal, or year-round and depends largely on the life-history requirements of each species. A restored wetland may be used simply as a source of water during a drought, as a migratory stopover, as overwintering habitat, or for breeding. Because so much wetland breeding habitat has been lost from the region, it is the evaluation of a restored wetland's function as breeding habitat that is most important. Surveys of animal breeding in restored wetlands indicate that within each group of animals some species are common breeders, some are infrequent breeders, and some have not been observed yet breeding in restored wetlands (Table 5.5). Many of the infrequent breeders have specialized habitat needs, such as a well-developed stand of a certain vegetation type or an adequate invertebrate food supply, and will be most likely to occur in restored wetlands that are similar to natural wetlands with these features. One simple way to evaluate the success of a restored wetland as breeding habitat is the number of species found in it that breed infrequently in restored wetlands. The more infrequent the breeders the better the wetland is functioning as wildlife habitat. If a species previously not observed breeding in restored wetlands is found, this is even a stronger indication that the wetland is functioning well as breeding habitat. For wetlands being restored for specified groups of animals, this breeding habitat evaluation can be restricted to just members of that group.

Except in cases where wetlands are being restored to create habitat for specific species, we believe that the simplest and most reliable way to evaluate a restored wetland as wildlife habitat is to compare the vegetation of a restored wetland to that of comparable natural wetlands. This avoids most of the uncertainties associated with animal recolonization and the censusing of restored wetlands that make the direct evaluation of wildlife usage so difficult to interpret.

CASE STUDY ────────────────────────────────

BIRD USE

Delphey (1991) compared bird use in restored wetlands with that in natural wetlands during the breeding season, late spring and early summer. The natural and restored semi-permanent wetlands of one pair that he sampled were approximately 1 mile apart in Dickinson County, Iowa. The natural wetland is a 4-acre basin (mean depth = 1.7 feet) within the Grovers Lake State Wildlife Area. The restored wetland is a 6-acre basin (mean depth = 1.4 feet) within the McBreen Marsh Wildlife Area. It was restored in 1988. Both areas are within the

TABLE 5.5. Species within each animal group observed or not yet observed breeding in restored potholes.

Group	Definition	Common Breeding Animals of Restored Wetlands	Infrequent Breeding Animals of Restored Wetlands	Not Observed in Restored Wetlands of the Region
AS	Birds with large area requirements - generally complexes of wetlands and associated grasslands.	none	none	Trumpeter swan (*Cygnus buccinator*), willet (*Catoptrophorus semipalmatus*), whooping crane (*Grus americana*), sandhill crane (*Grus canadensis*), long-billed curlew (*Numenius americanus*), marbled godwit (*Limosa fedoa*), northern harrier (*Circus cyaneus*), short-eared owl (*Asio flammeus*).
OW	Birds that require large, semipermanent wetlands or lakes. Many of the birds are colonial waterbirds or fish-eating species.	Pied-billed grebe (*Podilymbus podiceps*)	Black tern (*Chlidonias niger*)	Horned grebe (*Podiceps auritius*), eared grebe (*Podiceps nigricollis*), western grebe (*Aechmophorus occidentalis*), red-necked grebe (*Podiceps grisegena*), American white pelican (*Pelecanus erythrorhynchos*), great egret (*Casmerodius albus*), great blue heron (*Ardea herodias*), green-backed heron (*Butorides virescens*), black-crowned night heron (*Nycticorax nycticorax*), white-faced ibis (*Plegadis chihi*), redhead (*Aythya americana*), canvasback (*Aythya vallisneria*), ruddy duck (*Oxyura jamaicensis*), ring-necked duck (*Aythya collaris*), common loon (*Gavia immer*), Franklin's gull (*Larus pipixcan*), Forster's tern (*Sterna forsteri*).
MG	Marsh generalists: birds that can use smaller wetlands and require some robust emergent vegetation.	Red-winged blackbird (*Agelaius phoeniceus*), yellow-headed blackbird (*Xanthocephalus xanthocephalus*), American coot (*Fulica americana*).	none	Common moorhen (*Gallinula chloropus*).
SB	Secretive birds of shallow marshes including birds that require sedge meadows and wet prairie.	none	Common yellowthroat (*Geothlypis trichas*), least bittern (*Ixobrychus exilis*), Wilson's phalarope (*Phalaropus tricolor*), sedge wren (*Cistothorus platensis*), marsh wren (*Cistothorus palustris*), Virginia rail (*Rallus limicola*), sora (*Porzana carolina*), swamp sparrow (*Melospiza georginana*).	American bittern (*Botaurus lentiginosus*), king rail (*Rallus elegans*), LeConte's sparrow (*Ammospiza lecontei*), savannah sparrow (*Passerculus sandwichensis*), Henslow's sparrow (*Ammodramus henslowii*).
DG	Dabbling ducks and geese - often require several kinds of marshes to complete life stages.	Canada goose (*Branta canadensis*), mallard (*Anas platyrhynchos*), blue-winged teal (*Anas discors*).	Gadwall (*Anas strepera*), northern pintail (*Anas acuta*), green-winged teal (*Anas crecca*), northern shoveler (*Anas clypeata*).	
SH	Birds that require extensive bare soil - mud, sand, or gravel, for nesting and foraging - shorebirds.		Killdeer (*Charadrus vociferus*), spotted sandpiper (*Acititus macularia*)	Common snipe (*Capella gallinago*), American avocet (*Recurvirostra americana*), ca 25-30 species of migrating shorebirds.
RA	Reptiles and amphibians.	Tiger salamander (*Ambystoma tigrinum*), snapping turtle (*Chelydra serpentina*), American toad (*Bufo americanus*).	Northern leopard frog (*Rana pipiens*), bullfrog (*Rana catesbeiana*), chorus frog (*Pseudacris triseriata*)	Canadian toad (*Bufo hemiophrys*), cricket frog (*Acris crepitans*), western painted turtle (*Chrysemys picta*), blanding's turtle (*Emydoidea blandingi*), smooth green snake (*Opheodrys vernalis*), eastern garter snake (*Thamnophis sirtalis*), western plains garter snake (*Thamnophis radix*), mudpuppy (*Necturus maculosus*).
SM	Small mammals - mammals inhabiting wet prairies and sedge meadows.	no surveys	no surveys	Masked shrew (*Sorex cinereus*), pygmy shrew (*Microsorex hoyi*), short-tailed shrew (*Blarina brevicauda*) Franklin's ground squirrel (*Spermophilus franklinii*), meadow vole (*Microtus pennsylvanicus*), meadow jumping mouse (*Zapus hudsonius*), white-footed mouse (*Peromyscus leucopus*), deer mouse (*Peromyscus maniculatus*) southern bog lemming (*Synaptomys cooperi*).
FB	Fur bearers - mammals requiring semi-permanent marshes.	Muskrat (*Ondatra zibethicus*)	no surveys	Mink (*Mustela vison*), beaver (*Castor canadensis*), ermine (*Erminea bangsi*), long-tailed weasel (*Mustela frenata*), least weasel (*Mustela nivalis*).

same wetland complex and near several large lakes.

In spite of these similarities, their potential habitat suitability for birds was different. Both wetlands had open water (with submersed aquatics) and deep emergent zones. The natural marsh also had a continuous cover of vegetation in the shallow emergent and sedge meadow zones, whereas the restored basin had only a few scattered plants in those zones. A small area of wet prairie occurred in the natural basin, although this buffer was minimal because crops are planted close to the basin along one side and a road bounds the other side. No wet prairie was found in the restored marsh. Consequently, both sites are suitable for dabbling ducks and geese (Group DG) and birds that require emergent vegetation (Group MG). Neither site was suitable for birds that have large area requirements (Group AS and OW). Birds of sedge meadows and wet prairies (Group SB) should have found suitable habitat only at the natural marsh, and shorebirds (Group SH) would be likely to use the bare ground found along the margins of the restored marshes. Bird species that actually used both the restored and natural marshes (Table 5.6) were members of Group MG (marsh generalists) and Group DG (dabbling ducks and geese). The marsh wren, swamp sparrow, and common yellowthroat were absent from the restored marsh, presumably because of the lack of suitable habitat, sedge meadows. Shorebirds did not use mudflats in the restored marsh during the observation period. The sites were not visited during the peak migratory periods for shorebirds, in early spring and midsummer. It is very likely that some shorebirds did use the mudflats of the restored marsh during their migration.

These data suggest that management could be used to improve the restored marsh as wildlife habitat. The restored basin is attractive to dabbling ducks and geese but not so for sedge meadow species. These birds will not use the restored wetland until a sedge meadow develops. Unfortunately, there is no evidence that a sedge meadow will develop naturally, and it may be necessary to plant meadow species using plugs of vegetation and/or seed. Planting the wetland margin may reduce the attractiveness of this wetland for shorebirds, except during drawdowns. If not planted, the mudflats will become colonized by emergent species, such as cattails, or weedy species, such as reed canary grass or purple loosestrife, making the site unsuitable for shorebirds and sedge meadow/wet prairie birds.

TABLE 5.6. Birds observed on a restored and natural wetland in Dickinson County, Iowa

BIRD SPECIES	GROUP	NATURAL WETLAND	RESTORED WETLAND
Swamp sparrow	D	X	
Marsh wren	D	X	
Common yellowthroat	D	X	
Canada goose	E	X	X
Mallard	E	X	X
Blue-winged teal	E	X	X
Yellow-headed blackbird	C	X	X
Red-winged blackbird	C	X	X
Northern shoveler	E		X
Wood duck			X

Source: Delphey, 1991.

MANAGEMENT

Water control structures, inlets, and outlets require periodic maintenance. Frequent visits (especially during spring runoff) should help avoid
serious structural problems from developing. In particular, erosion rills
on dikes that result from wave and ice impact can become a major
problem if not promptly filled. Dike erosion should be checked each
spring as should damage by burrowing animals, particularly by muskrats. Appendix D8 is a data sheet that can be used to conduct maintenance inspections. All newly constructed sites should be visited several
times after periods of heavy precipitation to check the soundness of all
construction. The most common routine maintenance required is keep-
ing outlet structures free of debris. Algae and other debris frequently
clog trash racks and screens on the outlet after storms. Water levels then
may rise to the emergency spillway levels. If the spillway is not suffi-
ciently vegetated, severe erosion occurs. In many basins, outlets may
need to be cleaned every few weeks in late spring and early summer.
More than routine maintenance of dikes and structures, however, is normally needed for a successful restoration.

One of the most important benefits of regular evaluations of restored wetlands, particularly in the years immediately after restoration,
is that they will reveal whether or not any other problems have developed. These problems may be insoluble, e.g., expected hydrology does
not develop, but generally are soluble. Studies of wetland creation and
restoration have shown repeatedly that follow-up evaluations are an essential part of successful projects. Simply recreating the former hydrology of a site will not always result in the restoration of a wetland with

the composition, structure, and functions of a natural wetland. Many things can go wrong and frequently do. Conditions in the region have changed in so many ways that we cannot assume that wetlands being reestablished today will come to resemble those present before European settlement. In this section, we review problems that have occurred with the restoration of prairie wetlands and, where feasible, suggest solutions.

Water Regime

The most common water regime problem has been less water in the basin than anticipated. Regional lowering of the water table may prevent the basin from regaining its predrainage hydrology, particularly at sites with soil types (Histosols and Fluvaquents) that indicate these basins were formerly groundwater discharge areas. Occasionally, wetlands have been "restored" in areas that were formerly not wetlands. These are situations where little, if anything, can be done normally to salvage the project. Better site selection criteria that screen out such sites are the only solution. A lack of water may also indicate a structural defect. Missed tile lines leaving the site, a leaking bottom (in peaty or sandy areas), and leaking dikes are the three most common structural problems encountered in the prairie pothole region.

Sometimes, restored wetlands are flooded much longer and deeper than expected from basin and watershed characteristics because groundwater flow may supply water to the site that was not anticipated, undetected upslope tile lines may be discharging into the basin, precipitation may be above average, or there may have been errors in the watershed delineation. Standpipes or spillways can be lowered to remove additional water during storm runoff, if the additional water is undesirable. In extreme cases, water control structures may have to be altered to handle the higher than expected inputs. In a few extreme cases, dikes and water control structures have been washed away. Again, better site selection criteria are the only solution.

Revegetation

Occasionally a basin will not revegetate. This situation has been uncommon in prairie potholes and has been seen on only a few sites drained by tile for many years or that were excavated. More typically, the revegetation of a basin is partial. Some groups of species become established, e.g., submersed aquatics, while others do not, e.g., sedge meadow species. Likewise, the rates of revegetation differ from basin to

114

basin. It is also possible, but it has rarely happened, that rare or endangered plant species may become reestablished in a restored wetland. We first briefly review what is know about revegetation patterns, then identify specific problems that have been encountered, and finally briefly consider restorations where rare species occur. The active revegetation (seeding, planting, donor seed banks) of restored wetlands is taken up in the next chapter.

REVEGETATION PATTERNS

Some wetland species were often present in a basin before restoration as either seeds or adult plants. Mudflat annuals, including pinkweed (*Polygonum pensylvanicum*) and pigweed (*Amaranthus* spp.), are ubiquitous in restored basins. These species are present as weeds in agricultural fields. Some emergent plants, such as river bulrush and aquatic smartweed, also routinely occur in restored basins. These species seem to be able to persist in cultivated fields because of their tough rhizome systems. Wetlands that were drained with ditches rather than tiles often had emergent plants growing in these ditches that spread into the restored wetland when the ditches were plugged (Galatowitsch, 1993). Even in cultivated fields with tile systems, sporadic ponding occurs during periods of very high precipitation. Some fields with inadequate drainage may have ponded annually. This allows wetland species to become established periodically from seed and thus helps maintain these species in the seed bank. Relict wetland seed banks can play a role in the revegetation of restored wetlands, but they are only important typically in wetlands drained less than 20 years because of the death of seeds in the seed bank and its burial by incoming sediment. Erosion has added several feet of upland soil to many wetland basins in the prairie pothole region since drainage.

Restored wetlands revegetate at different rates and in different ways, depending on their past drainage and agricultural history. Rapid revegetation is expected for (1) pasture sites, (2) sites drained by tile for less than 20 years, and (3) sites drained with ditches. These sites often have good vegetative cover but can have low species diversity because aggressive colonizers like cattails cover most of the basin. Slow revegetation is expected for (1) sites drained with tiles for more than 20 years and (2) sites where most of the basin was excavated. Basins drained with tiles often initially only have a few submersed aquatics and many mudflat species in areas without standing water. Within a few years, more submersed aquatic species often become established, and emergents appear along the perimeter in narrow bands. Wet prairie and sedge meadow species, however, are seldom found in these wetlands.

REVEGETATION PROBLEMS

Revegetation problems in the prairie pothole region fall into two general categories: (1) unsuitable conditions for plant establishment and/or survival and (2) establishment and spread of undesirable species.

Unsuitable Conditions. The litter of cover crops on a site can create conditions unfavorable for establishment for all groups of wetland plants. This has been a problem on some sites previously in government set-aside programs. Heavy stands of fallen plant litter especially from improved switchgrass (Blackwell switchgrass) and smooth brome seem to impair wetland plant establishment. Plant litter in natural wetlands has been shown to reduce the total number of seeds that germinate (van der Valk, 1986). Seeds do not germinate because shading by a litter layer lowers soil temperatures in spring, and the litter is a physical barrier to seedling growth. Litter from some cover crops have been very slow to decompose in restored wetlands. In particular, switchgrass remains as standing and fallen litter for several years in newly restored wetlands. Basins with a thick litter layer from a cover crop will often benefit from a preconstruction burn to reduce the amount of standing and fallen litter and thus improve conditions for the recruitment of wetland species.

Excavated areas in restored wetlands often fail to revegetate. Exposed subsoil typically contains more clay and little or no organic matter when compared to topsoil. When it dries, it often becomes so hard that it prevents root penetration. At least two sites have been restored in the region where the entire basins were excavated to subsoil. After three years, the surface of each was still a very dense gray clay, and neither had revegetated. Whether this was because of unfavorable soil conditions for establishment or because all wetland seeds and propagules were removed from the basins is unknown. When feasible, excavated areas should be covered with a layer of topsoil.

A lack of vegetation may also indicate that the site has been flooded too deeply. Seeds of most emergent, sedge meadow, and wet prairie species will not germinate underwater. Seedlings of these species may also be killed if they are flooded too soon. Water levels may need to be lowered to facilitate seed germination at sites were a good wetland seed bank is likely to be present.

Submersed aquatic species sometimes have problems becoming established and often have problems surviving in restored potholes because of water quality problems. Excess turbidity is the most widespread water quality problem. It reduces the light penetrating the water column and can strongly influence the growth of submersed plants (e.g., Meyer and Heritage, 1941; Robel, 1961). For example, shaded plants may fail to produce tubers or seeds needed to overwinter (Kimber, 1994). Prairie potholes are naturally fertile wetlands and probably naturally had periodic algal blooms (Crumpton, 1989). Restored wetlands in agricultural landscapes often have frequent and severe algal blooms. Shading caused by algal blooms can adversely affect the growth of submersed plants and, in extreme cases, can eliminate them. Thick mats of filamentous algae (metaphyton), often nearly a foot thick, have developed on some recently restored prairie potholes. These thick mats severely restrict the growth of submersed aquatic plants.

Algal blooms are often controlled in farm ponds with copper sulfate or another algicide. However, these treatments are often expensive and short-lived, and so repeated doses of toxic chemicals must be applied to limit algal growth. Biological control methods for algae are beginning to be tested on a large scale, primarily in Minnesota and in Great Britain (Moss, 1990; Newbold, 1992). Maintaining large populations of zooplankton that feed on algae shows promise for controlling algal growth in shallow lakes and larger deep marshes. Large *Daphnia* species graze most effectively but are themselves a favored food source of planktivorous fish, such as panfish. Therefore, fish should not be stocked in newly restored marshes. No work has been done on biological manipulations in prairie potholes for the control of algal populations.

Turbidity problems can also be created by fine particles from the sediment being resuspended by wind-generated waves. This problem is most severe in restored wetlands that do not have well-established populations of submersed aquatic plants to anchor the sediment. Carp can reduce submersed aquatic plant populations by stirring up the bottoms of wetlands while feeding and breeding. Carp can easily reach basins connected to streams and ditches. But carp can also invade basins with only tile connections to streams or lakes because young carp travel through tile lines. Barriers to fish movement, such as screens on inflow and outflow pipes, should help reduce the potential for invasion. Unfortunately, such screens quickly clog with debris. Drawing down the basin is often the simplest way to kill carp. This should be done in fall, winter, or early spring to prevent invasion by cattail of exposed mudflats. Cattails can rapidly cover basins if they are drawn down in summer. Fall drawdowns should be done in late fall to minimize effects on native amphibians. Traditionally, the favored solution for carp eradication has been chemical treatment (Poff, 1985). Rotenone is applied immediately before ice cover to allow the chemical to take effect but detoxify before the wetland thaws. Unfortunately, the side effect of this treatment can be the eradication of leopard frogs and a severe reduction of invertebrate populations (Lannoo, 1993). The use of rotenone should be avoided.

Aggressive Weeds. There are six species (*Agropyron repens, Cirsium arvense, Lythrum salicaria, Myriophyllum spicatum, Phalaris arundinacea, Potamogeton crispus*) that are found in the region that grow so aggressively they can overrun an area and preclude other species from getting established or spreading. The presence of any of these species, even at low abundance, should be cause for intervention. Two other groups of species are often considered problems, cattails and woody plants. Whether or not they are indeed problems needs to be decided on a case-by-case basis and will be dependent largely on the goal of the restoration. The presence of cattails might be a problem for wetlands restored as wildlife habitat, but of no significance for wetlands being restored to improve water quality.

Eurasian milfoil (*Myriophyllum spicatum*) and curly-leaved pondweed (*Potamogeton crispus*) are submersed aquatics that are more common in lakes in the region but could potentially become problem species in semi-permanent and permanent restored wetlands. Both species may be controlled with drawdowns. Drawdowns for plant control should be done in the winter to avoid wildlife and recreation conflicts. At least three weeks of freezing and drying of soils are necessary to kill plants such as water milfoil.

Purple loosestrife (*Lythrum salicaria*) is an ornamental species from Europe that has escaped from gardens and infested shallow wetlands. It is a serious problem in some natural wetlands in the region and could become a serious problem in restored wetlands as well, although it is not yet. Purple loosestrife is known to invade moist, disturbed soils, conditions that are common in recently restored wetlands (Minnesota Department of Natural Resources, 1992). Although isolated wetlands are thought to be less susceptible to invasion since *Lythrum* seeds are water dispersed, its seeds are small enough to be carried in mud on animals and vehicles (Thompson et al., 1987; Thompson, 1989). Attempts at purple loosestrife control have been mostly futile because plants produce enormous numbers of seeds that remain viable in the sediments, plants resprout from roots and broken stems after being mowed, and adults can tolerate prolonged periods of flooding (Thompson et al., 1987; Gabor and Murkin, 1990). Several European insects are currently being evaluated as a biological control method for purple loosestrife. Early detection and eradication of invading plants with spot applications of herbicide (e.g., Rodeo-EPA Reg. No. 524-343) with hand-held sprayers is currently the best available control strategy (Thompson, 1989). Scattered young plants growing in soft sediments may be hand-pulled. Take care to remove all stem fragments and the root crowns of purple loosestrife from the wetland. Every effort should be made to prevent vehicles or other equipment from spreading it to restored wetlands from infested natural wetlands.

Reed canary grass (*Phalaris arundinacea*) is another weed of shallow marshes and can tolerate prolonged flooding. This species was promoted as a pasture species and for stabilizing the sides of ditches. Hand-digging plants can be effective to remove small patches (Diekelmann and Schuster, 1982). Large-scale mechanical removal may not be an adequate control technique because it quickly recolonizes from seed in the disturbed soil (Apfelbaum and Sams, 1987). Since reed canary grass grows in late spring and flowers in June, late spring burns may be effective, especially if coupled with active revegetation of desirable species. A two- to three-year burn rotation in an Illinois prairie successfully controlled reed canary grass (Apfelbaum and Rouffa, 1983). Herbicide applications may be necessary to control serious infestations of reed canary grass (Mason, 1992). Reed canary grass in restored wetlands is best controlled by applying herbicides directly to young plants before they set seed because it can spread quickly by seed (Comes et al., 1981). Glyphosate, Amitrol, and Dalapon have been used to control reed canary grass seedlings. Amitrol is effective on seedlings up to 3 weeks old while Glyphosate can be used to eradicate 5- to 10-week-old seedlings (Apfelbaum and Sams, 1987). In Indiana, Glyphosate and Dalapon have also been applied at flowering time to control reed canary grass.

Canada thistle (*Cirsium arvense*) and quackgrass (*Agropyron repens*) may establish in swales and on the periphery of restored wetlands. Thistle can be effectively controlled mechanically or with herbicides. Herbicide drift can have adverse effects on nontarget forbs in the wetland, and its use should be avoided, when possible. Prolonged flooding can also be used to kill these species, but because they are found only in the higher parts of the basin, this may only rarely be feasible.

Narrow leaf cattail and hybrid cattails are undoubtedly the most common and widespread wetland species within the region. Although they are common emergents in natural marshes, cattails are, nevertheless, often considered a problem in restorations because if they get established first it is difficult for other emergents, sedge meadow species, and wet prairie species to become established. Cattail seeds are dispersed widely by the wind and germinate readily on moist soil, especially when there is little competition from other species. Cattails can spread vegetatively and rapidly cover shallow marshes. Once established, their dense litter can also prevent other species from becoming established.

Since wind-dispersed cattail seeds are more common and widespread than seeds of other emergent species and because few wetland seeds of other emergents are normally found in cultivated basins, restored wetlands are often ideal sites for the establishment of cattails. Flooding the basin for several years might allow the seeds of animal-dispersed emergents to reach the site while reducing opportunities for the germination of cattail seeds. Cattails seeds, however, will germinate along the edges of flooded basins and can then invade flooded areas vegetatively.

Cattails have been controlled successfully by cutting their shoots and standing litter below water level to cut off the air supply to rhizomes and roots. Two to three repeated cuttings early in the season may be needed. Unfortunately, spring cuttings must be done underwater, which makes controlling large areas difficult. If winter water levels are lower than spring levels, standing litter can be mowed on the ice with a tractor and rotary mower during the winter. Any mechanical control will be ineffective near the periphery of a marsh. It may be necessary to combine mechanical control measures with active revegetation of other species. Preempting cattails with more desirable vegetation should reduce the potential for its reinvasion.

Today, many natural prairie potholes in the region are ringed with willows (*Salix* spp.) and cottonwoods (*Populus* spp.), unlike wetlands in more arid areas to the northwest. Perhaps woody invasion is a relatively recent phenomenon: presettlement prairie fires may have killed many of these trees, especially during droughts. Whether or not woody vegetation is undesirable around restored wetlands is a matter of opinion. Woody vegetation surrounding marshes makes ideal wildlife habitat for some species, such as woodpeckers and wood ducks. Trees, however, are also ideal perches for avian predators like crows and hawks. Willow seedlings will germinate on mudflats in newly restored wetlands. On some sites, these seedlings do not survive, while on others saplings will grow over a foot a year. Several cuttings of sapling stems should easily eradicate them. Burning can also be used to control shrub encroachment in sedge meadows. Burning can take place in August and September when woody species are still actively growing but after the nesting period of most marsh birds.

RARE PLANTS

It is possible and hopefully will be the case that some restorations may be colonized by rare wetland plant species. Although it is unlikely that recently restored wetlands will harbor orchids, bog beans (*Menyanthes trifoliata*), or wild rice, it is possible that such species may become established with time. The most likely group of rare species to be encountered in a restored wetland are fen species. Fen plants grow in areas where mineral-rich water discharges on the surface in areas not normally flooded. Some pasture sites being restored could potentially have suppressed populations of fen species. Since fens and sedge meadows have many species in common, the presence of a number of fen plants is needed to suggest that a fen may be redeveloping. Wetlands with rare species should be reported to state and federal agencies in charge of rare and endangered species programs. They should be carefully managed and protected from inputs of sediments, nutrients, and pesticides. Water level should be kept below historic elevations in the basin to avoid flooding fens. Herbicides also should not be used in these wetlands and surrounding buffers. A study of Iowa fens found that one-third of the sites had "edge-effects" from cultivation, including a lack of forbs because of broadleaf herbicide drift from nearby fields (Pearson and Leoschke, 1992).

Improving Wildlife Habitat

If evaluations of the use of restored wetlands by wildlife indicate that expected species are absent, there could be a problem with the restored wetland. The absence of a species, however, may have nothing to do with the features of the wetland that was restored. The species may simply not yet have reached the wetland. When problems with the restoration is the reason species are not using the wetland, it is possible to eliminate many of these problems. To determine if the absence of species is due to a problem with the restoration, water regime and vegetation data are normally required.

It is beyond the scope of this book to review all aspects of wildlife management in wetlands. This large and complex field is well covered in the book by Neil Payne (1992), *Techniques for Wildlife Habitat Management of Wetlands*. In this section, we examine the significance of the absence of species in natural wetlands (see Chapter 2) and suggest possible management strategies, if feasible, that could be used to make the restored wetland better habitat for missing species. In other words, the following information can be used to interpret the results of wildlife evaluations and to decide if there is a problem and what can be done to fix it.

AREA SENSITIVE BIRDS

Many birds included in this group have been extirpated from the region because they require hundreds to thousands of acres of prairie and wetlands. Restoration projects within a large area of suitable habitat may be used by one or more of these species, but only rarely, if ever. Some endangered animals, such as whooping cranes, may never again occur in this region. Any sightings of these species should be reported to your local wildlife officials.

Northern harriers may reestablish since they still breed sparingly within the region. Harriers hunt over large tracts of marshes and grassland, usually over several miles. Shallow marshes within areas of permanent cover may provide suitable habitat for northern harriers. Water levels should not exceed 6 inches during the nesting season (April–August) to prevent nest inundation (Hands et al., 1989b).

Trumpeter swans have been recently reintroduced to restored wetlands associated with Swan Lake in Nicollet County, Minnesota, and more reintroductions are planned for Minnesota and Iowa. Restored wetlands most suitable for trumpeter swans will be those greater than 40 acres that are surrounded by permanent grassland cover, are part of a wetland complex, and are free of rough fish such as carp (Andrews and Zenner, 1992). These wetlands will need to be protected from excessive agricultural runoff and carp invasion to permit the growth of submersed aquatic plants. Wetlands most attractive to swans will be those that are fairly isolated: not close to roads, buildings, or towns because trumpeter swans are thought to be sensitive to human disturbance. Since trumpeter swans also are sensitive to lead poisoning, restoration sites for this species should be those that have not historically been used intensively for hunting (Andrews and Zenner, 1992).

Marbled godwits need wetland complexes with a high diversity of wetland types (Ryan et al., 1984). Godwits tend to avoid wetlands with tall vegetation for feeding, opting for open to moderately vegetated shallow areas. Suitable upland habitat for godwit nesting are permanently vegetated areas, especially native grasslands and pastures. Godwits avoid tilled uplands. Like godwits, long-billed curlews and willets seem adapted to shorter prairie vegetation along wetland shores and in upland areas (Ryan and Renken, 1987; Ryan et al., 1984; Cole and Sharpe, 1976; Jenni et al., 1982). Formerly, periodic fire and bison grazing created large patches of short vegetation around wetlands. To reestablish suitable habitat for marbled godwits, willets, and long-billed curlews requires restoring a wetland complex that includes ephemeral and temporary wetlands; restoring surrounding upland areas to native grassland; moderate grazing by livestock in upland areas; and fall burning along wetland margins (Ryan and Renken, 1987; Ryan et al., 1984). Wetland margins should be burned in the fall, not spring, so the vegetation will still be short the following spring.

OPEN WATER BIRDS

Many birds that require open water to feed may use adjacent restored wetlands to nest. Shallow wetlands restored next to lakes (greater than 40 acres) could offer habitat for these open water birds. Although many of these species are not common in the region, some, like the pelican, are currently expanding their range. Pelicans are colonial and nest on the ground. Their colonies generally are on islands or isolated peninsulas where ground predators have little access to them. They will fly long distances from the colony to feeding sites, suggesting they have very specific nest site requirements. Others, such as the common loon and black tern, have solitary nests in well-developed emergent vegetation (Bergman et al., 1970). Sites with inadequate nesting cover may be less suitable for these species. Forster's terns prefer drier nest sites over water, such as nesting islands created by muskrat lodges (Weller and Spatcher, 1965).

Black terns may nest on old muskrat houses or may construct fragile nests that float on water in stands of emergent vegetation. The emergent plant material is a critical windbreak, without which nests are lost. Consequently, black tern nesting is somewhat cyclic on large semi-permanent marshes, favoring wetlands with adequate vegetative cover but also with adjacent open water. Marsh complexes can be managed so that some marshes always have suitable water-vegetation interspersion. Three to 4 feet of water is needed through the nesting season, from May to August. Some limited success has been achieved with artificial black tern nests. Nest platforms should be placed 30 to 50 feet apart, in groups of eight (Hands et al., 1989a). Artificial nest structures are most often used when placed in a marsh location where terns are already nesting (Hemesath, 1992).

MARSH GENERALISTS

This group of birds only requires some open water with standing vegetation. Some generalist birds that readily use restored marshes are the American coot, yellow-headed blackbird, and red-winged blackbird. No special management is normally needed to attract these species.

SEDGE MEADOW BIRDS AND SECRETIVE BIRDS OF SHALLOW MARSHES

Marshes with dense meadow and shallow emergent vegetation provide the necessary seclusion for bitterns and rails. These species feed on small aquatic animals such as frogs within dense vegetation and nest in densely vegetated sedge meadows. Emergent vegetation generally reestablishes within three years in restored wetlands. But, even if the basin offers adequate cover, birds may not be present without a good food base (Hands et al., 1989d). Wetlands in watersheds that are primarily in row crops may receive significant amounts of pesticides, reducing populations of invertebrates and amphibians. Optimal foraging depths for these species is less than 1 foot.

Sedge meadows in swales or along the periphery of larger marshes are required by several sparrows, wrens, and warblers. The lack of sedge meadows on restored wetlands appears to preclude these species from colonizing restored wetlands (Delphey, 1991). Planting sedge meadow species may be required to establish suitable habitat for these birds.

DABBLING DUCKS AND GEESE

Increasing the amount of suitable habitat for waterfowl will not necessarily result in increases in their populations. Some species, such as the mallard and blue-winged teal, are immediately attracted to restored wetlands while others, such as the gadwall and northern pintail, are less likely to use restored sites. The most effective strategies for using wetland restorations to improve waterfowl populations will require consideration of individual species requirements. Although the southern prairie pothole region is currently primarily a migration area for most species rather than a breeding area, the guidelines presented here address management for both breeding and migrating birds.

The amount of nest predation and brood mortality is often related to surrounding land use (e.g., Duebbert and Lokemoen, 1976; Higgins, 1977). The ideal ratio of upland grassland to wetland area in agricultural areas is unknown, but it should be at least 3 upland acres for each wetland acre. Large areas of permanently vegetated upland will allow ducks to disperse their nests, and this will reduce nest predation (e.g., Sudgen and Beyersbergen, 1986). If the surrounding land is used for hay production, hay harvesting should be delayed until after July 15 to reduce significant losses of nests and hens.

Artificial nests, primarily bales of straw, mailbox designs, or open baskets can be placed in restored wetlands to encourage nesting. They are generally inexpensive to construct and install but can only accommodate one nesting hen, unlike larger nesting islands. Zenner et al. (1992) provides specific guidelines for waterfowl nesting structures. Round hay bales, turned on end, are perhaps the most common nesting structure placed in restored wetlands. They need to be placed in relatively shallow areas of marshes (18–30 inches) and must be replaced every two to three years. Open baskets, made of plastic, metal, or wood, can be mounted on a pipe and installed in locations of restored basins that receive minimal ice rafting. Mallards and Canada geese nest in open baskets whereas other dabbling ducks seem to avoid their use at Union Slough National Wildlife Refuge (Fleskes and Klaas, 1991). Dabbling ducks will sometimes nest in horizontal cylinders (mailbox design) mounted on posts. Cylinders are constructed of wire or plastic and completely lined with grass.

SHOREBIRDS

Several shorebird species largely migrate through the middle of North America, especially in the spring. They include the Hudsonian godwit (*Limosa haemastica*), pectoral sandpiper (*Erolia melanotos*), short-billed dowitcher (*Limnodromus griseus*), white-rumped sandpiper (*Calidris fuscicollis*), and the few remaining Eskimo curlews (*Numenius borealis*) and lesser golden-plovers (*Pluvialis dominica*) (J. J. Dinsmore, Iowa State University, personal communication). These migrating shorebirds use sheetwater ponds that form on flat, barren fields during spring melt, the banks of rivers, and mudflats in semi-permanent marshes. Shorebirds also have been using mudflats in newly restored wetlands. These new mudflats will not persist more than a few years in most restored wetlands. Larger restored marshes can periodically be drawn down to create mudflats for shorebirds if they have the necessary water control structures. Opportunities to manage wetlands for shorebirds may be greatest in restored complexes that contain multiple semi-permanent wetlands. One basin can be drawn down to create a mudflat for shorebirds while other wetlands in the cluster can remain flooded for use by other wildlife species. Providing mudflat habitat is particularly important for the fall migration when sheetwater ponds are not available. This fall migratory period is surprisingly long: the first southbound birds arrive in the region the last week of June and the last leave in early November. The migratory peak occurs in mid to late August (Dinsmore, 1992).

REPTILES AND AMPHIBIANS

Frogs and tiger salamanders will quickly colonize restored wetlands. However, some wetlands are better than others at supporting their populations. Since tadpoles of all native amphibians within the region (except mudpuppies) grow rapidly and reach adulthood in a short period of time, temporary and seasonal wetlands are important breeding areas in all but drought years. Breeding success can be particularly high in these wetlands because fish predators are absent, as are bullfrogs. Bullfrogs, an introduced species, overwinter as tadpoles and so require semi-permanent or permanent bodies of water. Bullfrogs and fish exact such a high toll on native amphibians that temporary and seasonal wetlands are their only reliable breeding areas. Unfortunately, native amphibians must still use more permanent bodies of water during drought, severely reducing their populations (Lannoo, 1993).

Bullfrogs, frequently used as fishing bait, should not be introduced into restored wetlands because they can harm native populations of frogs by eating their larvae (Jackson, 1991). If bullfrogs colonize a restored wetland, they can best be removed by seining or by fall drawdown. Ponds should be drawn down in late summer and early fall to kill bullfrog tadpoles (Lannoo, 1993). Any chemical treatments, such as rotenone, should be avoided because desirable species will be adversely affected.

Vegetative cover is required by some species such as Blanding's turtles and chorus frogs. In contrast, cricket frogs prefer to forage on mudflats, escaping to open water when avoiding predators. Rooted aquatic vegetation facilitates egg laying in many amphibian species. Amphibian populations may not be supported in some marshes if pesticide runoff from adjacent farm fields reduces insect populations (their food supply) or causes reproductive failure. Mudpuppies, now extirpated from the region, may be an example of a species susceptible to contaminants in agricultural runoff (Christiansen and Bailey, 1991).

SMALL MAMMALS

Extensive development of sedge meadow vegetation and associated insect populations is required for small mammals. Birney and others (1976) suggested that some minimal quantity of vegetative cover is needed to sustain meadow vole populations. Planting sedges and wet prairie grasses may be required to provide adequate cover. After vegetation is established, do not burn or mow more often than every three years. If vegetation is removed more frequently, cover may be insufficient to support high densities of voles (Lemen and Clausen, 1984). In addition, pesticide runoff that reduces insect populations could reduce their food base.

FUR BEARERS

The muskrat is the most common of the fur bearers. Lack of emergent vegetation should reduce muskrat occupancy, but some have been observed in completely unvegetated recently restored wetlands that are near natural wetlands. Muskrats disperse readily and should reach newly restored wetlands, even when somewhat isolated from other wetlands (Errington, 1963). Muskrats will feed on submersed aquatics present in many newly restored wetlands although emergent vegetation, especially cattails, are preferred. Optimal wetlands for muskrats are semi-permanent marshes, those that do not go completely dry in midsummer in most years yet are shallow enough to allow continued growth or at least peri-

126

CHAPTER 5

odic rejuvenation of emergent vegetation (Errington, 1963). Muskrats
spend their winters on ice-sealed marshes, living entirely in open water
beneath the ice. Therefore, wintering habitat for muskrats will be in
wetlands with deep enough water so that the wetlands do not freeze to
the bottom. Lodges constructed by muskrats provide important nesting
sites for some birds including the Canada goose, black tern, and Forster's tern (Fritzell, 1989).

Predators such as mink and raccoons likewise will quickly discover
restorations. Raccoons and skunks actually use a variety of resources in
the agricultural landscape including old buildings for den sites and waste
grain for winter food. Mink, on the other hand, are much more closely
tied to wetland habitats, especially semi-permanent wetlands. Mink will
den in abandoned muskrat lodges and feed on a variety of birds, mammals, and even invertebrates (Fritzell, 1989). So the best habitat for
mink will be areas with a suitable prey base of muskrats and nesting
birds.

REFERENCES

Andrews, R. and G. G. Zenner. 1992. Trumpeter swan restoration proposal.
Unpublished report, September 28. Iowa Department of Natural Resources,
Clear Lake. 14 pp.

Apfelbaum, S.I. and A.S. Rouffa. 1983. James Woodworth Prairie Preserve: a
case history of the ecological monitoring programs. In *Proceedings of the
Eighth North American Prairie Conference*. Kalamazoo, Michigan.

Apfelbaum, S.I. and C.E. Sams. 1987. Ecology and control of reed canary
grass (*Phalaris arundinacea*). Natural Areas Journal 7(2): 69–74.

Bergman, R.D., P. Swain, and M.W. Weller. 1970. A comparative study of
nesting Forster's and black terns. Wilson Bulletin 82(4): 435–444.

Birney, E.C., W.E. Grant, and D.D. Baird. 1976. Importance of vegetative
cover to cycles of *Microtus* populations. Ecology 57: 1043–1051.

Brown, M. and J.J. Dinsmore. 1986. Implications of marsh size and isolation
for marsh bird management. Journal of Wildlife Management 50: 392–397.

Causton, D.R. 1988. Introduction to vegetation analysis. Unwin Hyman, Lon-
don, England. 342 pp.

Christiansen, J.L. and R.M. Bailey. 1991. *The Salamanders and Frogs of Iowa*.
Iowa Department of Natural Resources. Nongame Technical Series No. 3.
24 pp.

Cole, T. and R.S. Sharpe. 1976. The effects of grazing management on a
sandhills prairie community III. Breeding bird density and diversity. Proceedings of the Nebraska Academy of Science 86:12.

Comes, R.D., L.Y. Marquis, and A.D. Kelley. 1981. Response of three perennial grasses to dalapon, amitrol and glyphosate. Weed Science 29: 619–
621.

Crumpton, W.G. 1989. Algae in northern prairie wetlands. IN A.G. van der

Valk (Ed.). *Northern Prairie Wetlands.* Iowa State University Press, Ames.

Delphey, P.J. 1991. A comparison of the bird and aquatic macroinvertebrate communities between restored and natural Iowa prairie wetlands. M.S. Thesis, Iowa State University, Ames.

Diekelmann, J. and R. Schuster. 1982. *Natural Landscaping: Designing with Native Plant Communities.* McGraw-Hill Book Company, New York. 276 pp.

Dinsmore, J.J. 1992. Personal communication, animal ecology department, Iowa State University, Ames, Iowa.

Duebbert, H.F. and J.T. Lokemoen. 1976. Duck nesting in fields of undisturbed grass-legume cover. Journal of Wildlife Management 40: 39–49.

Errington, P. 1963. *Muskrat Populations.* Iowa State University Press, Ames. 183 pp.

Fleskes, J.P. and E.E. Klaas. 1991. Dabbling duck recruitment in relation to habitat and predators at Union Slough National Wildlife Refuge, Iowa. U.S. Fish and Wildlife Service, Fisheries and Wildlife Technical Report 32. 19 pp.

Fritzell, E. 1989. Mammals in prairie wetlands. IN A.G. van der Valk (Ed.). *Northern Prairie Wetlands.* Iowa State University Press, Ames.

Gabor, T.S. and H.R. Murkin. 1990. Effects of clipping purple loosestrife seedlings during a simulated wetland drawdown. Journal of Aquatic Plant Management 28: 98–100.

Galatowitsch, S.M. 1993. Site selection, design criteria, and performance assessment for wetland restorations in the prairie pothole region. Ph.D. Dissertation, Iowa State University, Ames.

Hands, H.M., R.D. Drobney, and M.R. Ryan. 1989a. Status of the black tern in the northcentral United States. Report prepared for the U.S. Fish and Wildlife Service, Twin Cities, Minnesota. 15 pp.

Hands, H.M., R.D. Drobney, and M.R. Ryan. 1989b. Status of the northern harrier in the northcentral United States. Report prepared for the U.S. Fish and Wildlife Service, Twin Cities, Minnesota. 18 pp.

Hands, H.M., R.D. Drobney, and M.R. Ryan. 1989c. Status of the common loon in the northcentral United States. Report prepared for the U.S. Fish and Wildlife Service, Twin Cities, Minnesota. 26 pp.

Hands, H.M., R.D. Drobney, and M.R. Ryan. 1989d. Status of the least bittern in the northcentral United States. Report prepared for the U.S. Fish and Wildlife Service, Twin Cities, Minnesota. 13 pp.

Harrison, C.V. 1978. *A Field Guide to the Nests, Eggs, and Nestlings of North American Birds.* Collins Publishers, Cleveland, Ohio.

Hemesath, L.M. 1991. Species richness and nest productivity of marsh birds on restored prairie potholes in northern Iowa. M.S. Thesis, Iowa State University, Ames.

Hemesath, L.M. 1992. Personal communication, Iowa Department of Natural Resources, Boone, Iowa.

Higgins, K.F. 1977. Duck nesting in intensively farmed areas of North Dakota. Journal of Wildlife Management 41: 232–242.

Jackson, L.S. 1991. Amphibian Management: what you can do for frogs, toads, and salamanders. IN J.L. Christiansen, and R.M. Bailey (Eds.). *The Salamanders and Frogs of Iowa.* Iowa Department of Natural Resources Nongame Technical Series No. 3.

128

Jenni, D.A., R.L. Redmond, and T.K. Bicak. 1982. Behavioral ecology and habitat relationships of long-billed curlews in western Idaho. U.S. Bureau of Land Management. Boise, Idaho. 234 pp.

Kantrud, H.A. 1986. Effects of vegetation manipulation on breeding waterfowl in prairie wetlands—a literature review. U.S. Fish and Wildlife Service, Fisheries and Wildlife Technical Report 3. 15 pp.

Keeney, D.R. and T.H. DeLuca. 1993. Des Moines River nitrate in relation to watershed agricultural practices: 1945 versus 1980s. Journal of Environmental Quality. 22: 267–272.

Kimber, A. 1994. Potential for *Vallisneria americana* restoration in the upper Mississippi River. Ph.D. Dissertation, Iowa State University, Ames.

Klett, A.T., H.F. Duebbert, C.A. Faanes, and K.F. Higgins. 1986. Techniques for studying nest success of ducks in upland habitats in the prairie pothole region. U.S. Fish and Wildlife Service Resource Publication 158. 21 pp.

LaGrange, T.G. and J.J. Dinsmore. 1989. Plant and animal community responses to restored Iowa wetlands. Prairie Naturalist 21: 39–48.

Lannoo, M.J. 1993. Personal communication, Ball State University, Muncie, Indiana.

Lemen, C.A. and M.K. Clausen. 1984. The effects of mowing on the rodent community of a native tall grass prairie in eastern Nebraska. Prairie Naturalist 16: 5–10.

Linde, A.F. 1985. Vegetation management in water impoundments: alternatives and supplements to water-level control. IN *Water Impoundments for Wildlife: A Habitat Management Workshop*. U.S. Department of Agriculture. General Technical Report NC-100.

Ludwig, J.A. and J.F. Reynolds. 1988. Statistical ecology. Wiley Interscience, New York. 337 pp.

Mason, D. 1992. Personal communication, letter from Wetlands Research, Wadsworth, Illinois, November 20.

Meyer, B.C. and A.C. Heritage. 1941. Effect of turbidity and depth of immersion on apparent photosynthesis in submerged vascular aquatics. Ecology 24: 393–399.

Minnesota Department of Natural Resources. 1992. A field guide to aquatic exotic plants and animals. Exotic Species Program, St. Paul, Minnesota. 2 pp.

Moss, B. 1990. Engineering and biological approaches to the restoration from eutrophication of shallow lakes in which aquatic plant communities are important components. Hydrobiologia 200/201: 367–377.

Newbold, C. 1992. Macrophytes, algae and zooplankton—their interrelated role in restoration ecology. Presentation at INTECOL's IV International Wetlands Conference. September 13–18, Ohio State University, Columbus.

Payne, N. F. 1992. *Techniques for Wildlife Habitat Management of Wetlands*. McGraw-Hill, New York. 549 pp.

Pearson, J. and M. Loeschke. 1992. Floristic composition and conservation status of fens in Iowa. Journal of the Iowa Academy of Science 99(2–3): 41–52.

Poff, R.J. 1985. Managing waterfowl impoundments for fisheries. IN M.P. Knighton (Ed.). Water Impoundments for Wildlife: A Habitat Management Workshop. U.S. Forest Service General Technical Report NC-100.

Reid, F.A., W.D. Rundle, and M.W. Sayre. 1983. Shorebird migration chronology at two Mississippi River Valley wetlands of Missouri. Transactions of the Missouri Academy of Science 17: 103–115.

Robel, R.J. 1961. Water depth and turbidity in relation to growth of sago pondweed. Journal of Wildlife Management 25: 436–438.

Roberts, T.S. 1932. *Birds of Minnesota*. University of Minnesota, Minneapolis.

Ryan, M.R. and R.B. Renken. 1987. Habitat use by breeding willets in the northern Great Plains. Wilson Bulletin 99(2): 175–189.

Ryan, M.R., R.B. Renken, and J.J. Dinsmore. 1984. Marbled godwit habitat selection in the northern prairie region. Journal of Wildlife Management 48: 1206–1218.

Sudgen, L.G. and G.W. Beyersbergen. 1986. Effect of density and concealment on American crow predation of simulated duck nests. Journal of Wildlife Management 50: 9–14.

Thompson, D.Q. 1989. Control of purple loosestrife. Fish and Widlife Service Leaflet 13.4.11. 6 pp.

Thompson, D.Q., R.L. Stuckey, and E.B. Thompson. 1987. Spread, impact, and control of purple loosestrife (*Lythrum salicaria*) in North American wetlands. U.S. Fish and Wildlife Service, Fisheries and Wildlife Research 2. 55 pp.

van der Valk, A.G. 1986. The impact of litter and annual plants on recruitment from the seed bank of the lacustrine wetland. Aquatic Botany 24: 13–26.

van der Valk, A.G. and C.B. Davis. 1978. The role of seed banks in the vegetation dynamics of prairie glacial marshes. Ecology 59: 322–335.

van der Valk, A.G. and R.L. Pederson. 1989. Seed banks and the management and restoration of natural vegetation. IN M.A. Leck, V.T. Parker, and R.L. Simpson (Eds.). *Ecology of Soil Seed Banks*. Academic Press, San Diego.

Wienhold, C.E. and A.G. van der Valk. 1989. The impact of duration of drainage on the seed banks of northern prairie wetlands. Canadian Journal of Botany 67: 1878–1884.

Weller, M.W. 1975. Studies of cattail in relation to management for marsh wildlife. Iowa State Journal of Research 49: 383–412.

Weller, M.W. 1979. Birds of some Iowa wetlands in relation to concepts of faunal preservation. Proceedings of the Iowa Academy of Science 86: 81–88.

Weller, M.W. and C.E. Spatcher. 1965. The role of habitat in the distribution and abundance of marsh birds. Iowa State University Agriculture and Home Economics Experiment Station Special Report No. 43.

Zenner, G.G., T.G. LaGrange, and A.W. Hancock. 1992. Nest structures for ducks and geese: a guide to the construction, placement, and maintenance of nest structures for Canada geese, mallards, and wood ducks. Iowa Department of Natural Resources, Des Moines. 34 pp.

ACTIVE REVEGETATION

Reports from prairie pothole restorations in the mid-1980s suggested that restored basins became recolonized with plants within a year after reflooding. Consequently, most specifications developed for prairie pothole restorations have focused on construction techniques. In contrast, other kinds of wetland restorations across the United States, including riparian forests, coastal marshes, and sedge meadows, have required either seeding or planting of seedlings or rootstocks of most species found in extant examples of these wetlands to revegetate them. The construction phase of these projects is generally only the first phase of the restoration. Establishing suitable vegetation can take several years and involves finding a source of plant materials, adding soil amendments (organic matter, fertilizers), establishing species from seed or transplants, controlling undesirable species, and manipulating water levels and other environmental conditions to promote the growth of seedlings and transplants. The large numbers of prairie pothole restorations conducted since 1987 were possible, in large part, because active revegetation was not attempted.

It is theoretically possible that the vegetation of restored basins will eventually come to be like that of natural marshes—a result of normal dispersal of seeds and other propagules. How many years this will take is not known. It could take a very long time. What we do know is that this has not yet happened. This may simply be because most restorations are relatively recent (since the late 1980s), and there has not been adequate time. There are also suggestions, however, from studies of the vegetation of restored wetlands that there could be vegetation types for which it will never happen, at least for isolated restored wetlands in agricultural landscapes. It is unclear whether or not dispersal mechanisms operating today are comparable to those that operated prior to European settlement, particularly dispersal by water.

There are a number of species that are often found in restored wetlands. Submersed aquatic species, such as pondweeds, bladderwort, and duckweeds, appear to reestablish very rapidly on nearly all sites, probably at least in part due to waterfowl dispersal. Emergent aquatics, such as cattails and bulrushes, also seem to establish rapidly in most restorations. Sedge meadow species and wet prairie species, in contrast, are uncommon (Galatowitsch, 1993). Sedge meadows and wet prairies may very well have to be reestablished by active revegetation programs. If these vegetation types are not established early in a restoration, undesirable species, e.g., reed canary grass and cattails, may colonize and make it very difficult or impossible for wet prairie and sedge meadow species to be established later.

Guidelines and techniques for active revegetation are provided primarily for establishing sedge meadow and wet prairie vegetation. Some information is provided in passing on establishing submersed and emergent species. Restorations most likely to benefit from active revegetation are those that were thoroughly drained with tiles and cultivated for more than 20 years and those that were excavated to deepen their basins. The revegetation of these sites should begin as early possible, during the first year if weather conditions are suitable.

Establishing species in a wetland can been done in five basic ways: (1) using donor soil with its seed and roots, i.e., soil from an existing wetland is spread over the restored basin; (2) inoculating a restored basin with donor soil; (3) mechanical or hand seeding; (4) using wild hay; and (5) transplanting seedlings, rootstocks, or whole plants. Generally, the most effective revegetation programs will involve a combination of techniques because of the varying establishment requirements of different species. The two most promising techniques for prairie potholes are inoculation with donor soils and using wild hay because they are the easiest and the cheapest.

Plant material for revegetating basins should be obtained as close to the site as possible. This ensures that the plants are adapted to local environmental conditions and that genetic diversity of wetland species is preserved. Some wetland species are known to vary in their environmental tolerance from region to region. For example, coastal populations of cattails are more salt tolerant than inland populations. From region to region, cattails also vary in the rate of flowering and standing crops of roots and shoots (McNaughton, 1966). All wetland species that have been studied are composed of many different ecotypes adapted to local conditions. Unfortunately, very limited supplies of wetland plant materials are commercially available locally, and finding a source of local plant material may be very difficult or impossible. Appendix E includes a list of plant material sources, by species, for the southern prairie pothole region.

DONOR SEED BANKS

Occasionally, a situation arises where a natural wetland is being destroyed near a wetland being restored. Surface sediments from the natural wetland can be moved to the restored basin and spread across its entire surface. The donor soil contains seeds, roots, and other plant parts with nutrients and an organic-rich substrate often missing from the basin to be restored. Most seeds are contained in the top 2 to 4 inches of sediment and roots may extend down another 5 inches. So it only makes sense to scrape the top 8 to 10 inches of sediment from the natural marsh. Removing the sediment to a greater depth may dilute the number of seeds and roots per unit area when it is spread across the restored wetland. Focusing on the top 8 to 10 inches also reduces the need to remove soil from the receiving basin to make room for the donor soil. The sediment can be moved by front-end loader, transported by dump truck, and smoothed in place at the receiving wetland with a small dozer and scraper. In some cases, for instance sedge meadows, it may be most logical to remove the turf intact and lay it like sod on the restored site. This has been done in a number of wetland restorations in the Midwest.

The donor wetland basin should be field checked before soil is removed to ensure there are no troublesome weeds, such as purple loosestrife, that would be transported to the new site. Seeds of undesirable species may be present in the seed bank but not in the standing vegetation. If a problem is possible, a seed bank survey (Appendix D5) will detect the presence of unwanted species.

INOCULATING WITH DONOR SOIL

Situations where an entire restored basin or even a significant portion of the basin can be covered with donor soil will be very rare. However, using small amounts of donor soil to inoculate a restored basin should be routinely feasible. Usually only a few seeds at a time of most species will reach a restored basin as a result of normal seed dispersal mechanisms. Once established, species spread either vegetatively or by seed produced by the pioneer populations. Small amounts of donor soils from natural wetlands can be collected and spread over one or more small areas in a restored wetland. Soil from all the vegetation zones in natural wetlands that are expected to develop in the restored wetland should be spread in appropriate parts of the restored wetland. As little as a few cubic feet of donor soil could be sufficient for this purpose. Soil for inoculating a restored wetland can be collected from several different wetlands in the area. Because of changing environmental conditions that will favor the establishment of some species over others in any given year, inoculations should be done for a number of years. It is known from seed bank studies of natural wetlands that not all species present in natural wetlands are found in their seed banks. One group of species whose seeds are often not found in seed banks are sedges. Collecting a small amount of soil from a nearby natural wetland should do very little or no permanent damage to the donor wetland.

SEEDING

Most restorations that require active establishment of species are seeded because this is less expensive than transplanting seedlings or adults. Although seeding has not been tried to any extent in prairie pothole restorations, this technique has been used in restoration of other types of wetlands and most upland vegetation types. It is probably the only feasible technique for revegetating large numbers of sites. Also, since the greatest need for revegetation is for sedge meadow and wet prairie zones, most methods developed over the past 25 years for prairie restorations are appropriate for reestablishing these two zones. Recommendations that follow are based on those developed by Robert Betz of Northern Illinois University (Betz, 1980) and Peter Schramm of Knox College in Illinois (Schramm, 1978, 1992) for restoring tallgrass prairies.

Because the species that could be planted range from those that grow and spread quickly to those that are slow growing and often sensitive to overcrowding by other plants, first planting more aggressive species and then slower growing species is more successful than trying to establish them together. Since seed availability varies, growing conditions can be unpredictable, and seed can be expensive, and it is often advantageous to add species periodically to a restoration rather than trying to revegetate in one planting.

The first species planted should be those most capable of competing with aggressive weeds, such as reed canary grass and thistles. Stage one species for wet prairies and sedge meadows are listed in Table 6.1. The initial "matrix mix" should include late-flowering grasses — including cord grass (*Spartina pectinata*) — together with aggressive prairie forbs — coreopsis (*Coreopsis* spp.), tick trefoil (*Desmodium* spp.), coneflower (*Ratibida* spp.), and compass plants (*Silphium* spp.) — whose seed can be mechanically planted. This type of first stage mix has proved capable of suppressing reed canary grass and thistles at the Fermi Lab in northern Illinois. Some commercial ecotypes — especially of switchgrass — should not be used in the matrix mix because they are too vigorous and will make it difficult to establish other species.

A wet prairie–sedge meadow sward can be burned a year after planting. Forbs and grasses that can reestablish once the weedy vegetation has been reduced are then hand-sown, generally after two to three years when annual weed species are in decline. Stage two species are listed in Table 6.1. Prairie grasses then begin to expand and some showy cup plants and sunflowers should be obvious. While the dominant wet prairie species, namely grasses, can be seeded in the first or second stage of planting, sedges common in wetter areas cannot.

A third species, stage three (Table 6.1), cannot be seeded into wetlands because they are sensitive to competition from weeds and even other native species, especially grasses. These include sedges, gentians, orchids, and some lilies. They should be grown in flats and then transferred as seedlings to pots for a season. Second-year plants can be transplanted into wet prairies that are well established, i.e., five or six years after initial seeding. Forbs are scattered and in low abundance throughout natural wetlands, but usually in patches. Planting patches of forbs (rather than single plants) should help reduce competition from grasses. Many of these species are uncommon in natural wetlands, so existing populations should not be used as donor populations to restored wetlands. However, salvage operations are occasionally undertaken, and restored wetlands may be appropriate sites for those plants.

Seed can be purchased from nurseries that specialize in prairie and wetland plant materials (Appendix E1). Large commercial nurseries in the Great Plains sell prairie grasses, such as big bluestem and switchgrass, which are southern improved varieties, quite different in growth habits from local plant material. These species, namely Blackwell switchgrass, grow vigorously and so are ideal for soil conservation but not for wetland restoration projects. Seed can also be harvested mechanically or by hand. Some meadow species, such as sedges, will have ripe seeds by July, while many emergent and submersed species will not be harvestable until August or September. Mechanical harvesting has lim-

TABLE 6.1. **Suggested species for various stages of revegetation. Stage 1 plants (unless otherwise noted) can be mechanically seeded, Stage 2 plants are hand-seeded, and Stage 3 plants are transplanted as seedlings or plugs. Scientific names are provided in Appendix A.**

Group	Plants Often Recolonizing Without Planting	Stage 1 Plants	Stage 2 Plants	Stage 3 Plants	Weedy Plants To Be Avoided
WP - Wet prairie plants	Fleabane	Big bluestem Switchgrass Coreopsis Tick clover Gray coneflower Cup plant Compass plant	Prairie phlox Culver's root Golden alexander Blazing star Mountain mint Canada anemone	Prairie gentian Yellow star grass Michigan lily Wood lily Heavy sedge	Quackgrass Blackwell switchgrass
SM - Sedge meadow plants	Fox sedge Rice cut grass Dudley's rush Torrey's rush Beggar's ticks Curly dock	Cord grass (plugs) Boneset Joe pye weed Woundwort Monkey flower Blue vervain Asters	Bluejoint Swamp milkweed Skullcaps Bugleweeds Yellow loosestrifes	Woolly sedge Tussock sedge Fringed gentian Bottle gentian Western fringed prairie orchid White ladyslipper	Canadian thistle Sawtooth sunflower
SE - Shallow emergent plants	River bulrush Water plantain Broad leaf cattail Wapato Giant burreed Water smartweed	American sloughgrass Tall manna grass White top	Water parsnip Blue flag Tufted loosestrife	Awned sedge Lake sedge Beaked sedge	Reed canary grass Purple loosestrife

DE - Deep emergent plants	Hard-stemmed
	  bulrush
	Soft-stemmed
	  bulrush
		none	none	Wild rice	Narrow leaf cattail
					Hybrid cattail

SA - Submersed aquatic plants	Coontail
	Common
	  bladderwort
	Bushy pondweed
	Sago pondweed
	Leafy pondweed
	Baby pondweed
	Flat-stemmed
	  pondweed
	Long-lived
	  pondweed
			Other pondweeds
			Water stargrass
			Crowfoots
				Wild celery	European milfoil
					Curly muckweed

FA - Floating annual plants	Lesser duckweed
	Greater duckweed
	Star duckweed
		none	Purple-fringed
			  riccia
			Slender riccia
			  (whole plants)
				none	none

MA - Mudflat annual plants	Most plants including: Nutgrasses Annual spike rushes Barnyard grass Pigweeds Beggar's ticks Smartweeds				
WO - Woody plants	Sandbar willows Cottonwood				

Source: Plant material information based on Schramm, 1990, and Betz, 1980.

ited application when harvesting seed from natural wetlands because seed ripening is variable and because some species are only in flooded areas. Front-mounted combines have worked well for prairie seed harvest in the Midwest.

Prairie grass and forb seeds typically must be exposed to cold, damp conditions (vernalization or stratification) for maximum and rapid germination. Moist or dampened seeds should be stored above freezing immediately after collection. Seed stored too dry may lose viability. To dampen seeds, add 1 part water to 50 parts seed and store just above freezing in covered plastic containers or bags. Do not add sand or vermiculite if seed is to be planted mechanically because it fouls planting devices. Most seeds should be stratified six to eight weeks, although very small seeds may require only a week or two of stratification. Grass seed should be allowed to air dry for a day or two to facilitate hand-seeding or mechanical spreading. Forb seeds do not need to be air dried before spreading.

Seeds are best planted in the spring. Seeds planted in the fall are going to be more susceptible to frost-heave and being eaten by rodents. Weed control is also less effective in the fall. Site preparation of wetland perimeters can begin as water levels recede in late summer in the year before planting. Mow the existing vegetation to a height of less than 1 foot and plow to a depth of 9 inches during the summer before planting. Freezing and thawing during the intervening winter will kill some exposed weed roots. Disk the area in the spring, when the soil is dry enough to work. Harrow several times before seeding to level the area and dislodge existing roots. Lightly cultivate the area to be seeded until mid-June to eliminate weeds as they germinate. Deep cultivation will stir up more weeds. Finally, use a cultipacker to make a firm base for seed application. Broadcast the seed mix with an all-terrain spreader or fertilizer applicator and cultipack again. An all-terrain spreader is a modified salt truck that surface broadcasts the seed. The postseeding pass with the cultipacker ensures adequate contact of the seed with the soil. Some difficulty can arise when broadcasting different kinds of seed together. Heavy seeds will tend to settle in the bin, resulting in an uneven application of different species.

Approximately 20 pounds per acre of a mix of grasses and forbs seeds should be planted. An efficient use of seed may be to spread seeds of widespread species in one pass over the entire area with subsequent passes over a smaller portion of the site, with more specialized species (Jacobsen, 1992). This permits a greater diversity of seeds to be planted without wasting seed. Some forb-rich patches can be planted that only include 1 pound per acre of grass seed, whereas the remaining areas may have up to 15 pounds per acre of grasses.

Shallow seed drills (such as Nesbit or Truax rangeland grass drills) and hydromulchers have also been used to plant prairie and wetland seed. Use of seed drills was discontinued at the Fermi Lab prairie in northern Illinois because planted rows are visible even a decade after seeding. Seed drills also require very clean and dry seeds to operate correctly. Hydromulchers use a slurry of seeds in water to spray onto prepared soil. The water tanks make hydromulchers very heavy, and they easily bog down in wet soil. The all-terrain spreader in use at the Fermi Lab is a salt spreader with low-pressure tires. County Conservation Boards or Soil and Water Conservation District Offices occasionally have equipment to loan for seeding prairies.

Weedy annual plants such as pigweed and foxtail will be most obvious during the first year. Native planted species should be evident the second year. Observations from prairie restorations in Illinois suggest that revegetation will be enhanced by annual fires for the first 5 years. Annual, early spring burns (early March through April) clear away accumulated litter and allow the soil to warm earlier. Prairie plants that begin to grow earlier get a head start on weeds. Early spring burns leave winter cover for wildlife and avoid the upcoming nesting period for many birds. Half the site can be burned for the next 5 years—with the other half set aside as a refuge for insects and other animals sensitive to fire. After 10 years, burns every 3 to 4 years should be adequate. Mowing and grazing are not substitutes for fire because both remove live and dead standing vegetation, often during the growing season.

WILD HAY

There is another approach to seeding restored wetlands that has been used to some extent with prairie restorations, but it seems not to have been tried with wetland restorations: wild hay. Wild hay is collected in a natural wetland in the late summer or fall when seeds of most species are ripe. For sedge meadows, hay may have to be collected in midsummer, because sedges flower in early summer. This hay is then spread on the appropriate portions of the restored wetland. Wild hay can be collected from different local marshes to apply to a particular restored wetland. Several applications of wild hay could also be made over a period of years. There are years when some species in natural wetlands flower more than others. Applications of wild hay from different marshes over a number of years should greatly improve the species diversity in the restored wetland. Sedge meadow and wet prairie species are the best candidates for this method of establishing vegetation. This method allows local seed sources to be used, introduces seeds of many species simultaneously, and is cheaper and easier than collecting, cleaning, storing, and planting seeds of these species

TRANSPLANTING

Some wet meadow plants, such as sedges, are known to be difficult to establish from seed directly. These stage three species are probably best revegetated after being reared for up to two years in pots. Seeds are planted in flats or in small pots and maintained in greenhouse conditions, with adequate light, heat, and water. Adequate root mass is critical for these plants to survive transplantation. A Wisconsin nursery has found that plants with an inadequate root system are easily washed out (Kerans, 1990). For sedges, two years of growth provides the needed root mass to ensure anchoring. Sedges vary in their sensitivity to handling: for example, awned sedge (*Carex atherodes*) and lake sedge (*Carex lacustris*) must be planted quickly after being extracted from the nursery wetland or greenhouse flats, whereas tussock sedge (*Carex stricta*) can tolerate some delay during the harvesting-planting process (Mason, 1992).

Donor wetland sites have also been used to extract roots and tubers or turf plugs, rather than excavating the entire surface of the basin. Roots and plugs can often be extracted with minimal disturbance to the wetland. Plugs of turf may be an appropriate way to revegetate wet meadow areas. Individual root sections of sedges and rushes are small. Plug transplants have been successful in establishing coastal marshes. Species that grow rapidly from underground roots and stems are candidates for root transplants: bugleweed (*Lycopus* spp.), arrowhead (*Sagittaria* spp.), wild celery, flag grass (*Phragmites* spp.), sweet flag (*Acorus* spp.), spike rush (*Eleocharis* spp.), pickerelweed (*Pontederia cordata*), bulrushes, cattails, waterlilies (*Nuphar* and *Nymphaea*), and pondweeds (*Potamogeton* spp.). Rhizomes can be collected while plants are dormant; that is, from October through April in the prairie pothole region. In general, early spring plantings are most successful. In Illinois, sedge rhizomes transplanted in March had a 90% survival rate, whereas those transplanted in May had a survival rate of less than 10% (Mason, 1992). However, burreed (*Sparganium* spp.), dock (*Rumex* spp.), bulrush, rush (*Juncus* spp.), and arrowhead (*Sagittaria* spp.) have been successfully planted in the fall. Wetland plant roots often grow in dense mats and will need to be dug with a spade or sharp knife. Rhizomes should be kept wet to moist after being dug up. Root materials must be planted deep enough in the soil so they do not float out. The spacing of planting is dependent on growth rate. Fast growing species such as river bulrush can be planted at 4–5-foot intervals, whereas slower growing species such as hard-stem bulrush should be planted at 2-foot intervals.

Submersed aquatic plants often readily become established in restored wetlands. However, at some sites more desirable species, such as sago pondweed (*Potamogeton pectinatus*) and wild celery have not become established. These latter two species are best established from rootstock (tubers) after adequate site preparation. Establish several patches of submersed aquatics in relatively sheltered portions of the basin with a full pool depth of 2.5 and 3.5 feet deep. Somewhat sandy soils will be more suitable than hard clay soils (Korschgen and Green, 1988). Select areas with minimal disturbance from wave action and ice pileup. Collect or obtain tubers no more than two weeks before planting. Keep the tubers cool (refrigerated) and moist until planting. Plant in midspring (May), before the tubers sprout. Tubers should be planted at a depth of 2 to 4 inches (Korschgen and Green, 1988). Mixed plantings, for instance of sago pondweed and wild celery, may actually be more successful than monotypic stands because their different growth forms will be complementary in reducing water turbulence in the stand (Kimber, 1992). Plant tubers approximately 2.5 feet apart. If algal mats are a problem, mechanically remove algae from the area prior to planting.

Exclosures of chicken wire and fence posts can be used to protect young plants from rough fish and act as a windbreak (if possible, remove rough fish before restoring submersed aquatics). Each exclosure should be approximately 6 feet by 12 feet, with the long axis along the contour of the basin. Reestablished beds that are too small may not sustain themselves because they may be ineffective at stabilizing sediments and reducing turbidity (Carter and Rybicki, 1985). The top of the exclosure should be above the water surface. Exclosures are also often needed to protect emergent, sedge meadow, and wet prairie seedlings from herbivores, especially geese.

Water-level management is important during establishment of species from rhizomes, seeds, or transplanted seedlings. Flooding too deeply can uproot plants and can kill seedlings. Many plants such as spike rushes and rushes can withstand low soil moisture conditions while other species such as manna grass (*Glyceria* spp.) cannot. Submersed species need to be continuously submersed during the first growing season. Emergent species should be shallowly flooded.

REFERENCES

Betz, R. 1980. One decade of research in prairie restoration at the Fermi National Acceleration Laboratory, Batavia, Illinois. Proceedings of the Midwest Prairie Conference 9: 179–185.

140

Carter, V. and N.B. Rybicki. 1985. The effects of grazers and light penetration on the survival of transplants of *Vallisneria americana* in the tidal Potomac River, Maryland. Aquatic Botany 23: 197–213.

Galatowitsch, S.M. 1993. Site selection, design criteria, and performance assessment for wetland restorations in the prairie pothole region. Ph.D. Dissertation, Iowa State University, Ames.

Jacobsen, E. 1992. Personal communication, U.S. Soil Conservation Service, Midwest Technical Center, Lincoln, Nebraska.

Kerans, K. 1990. The restoration business. Restoration and Management Notes 8(1): 29–31.

Kimber, A. 1992. Personal communication, botany department, Iowa State University, Ames.

Korschgen, C.E. and W.L. Green. 1988. American Wildcelery (*Vallisneria americana*): Ecological Considerations for Restoration. U.S.Fish and Wildlife Service Technical Report 19. 24 pp.

McNaughton, S.J. 1966. Ecotype function in the *Typha* community type. Ecological Monographs 36: 297–325.

Mason, D. 1992. Personal communication, Wetlands Research, Wadsworth, Illinois.

Schramm, P. 1978. The "do's and don'ts" of prairie restoration. IN D.C. Glenn-Lewin and R.Q. Landers, Jr. (Eds.). *Fifth Midwest Prairie Conference Proceedings*. Iowa State University, Ames. pp. 139–150.

Schramm, P. 1990. Prairie restoration: a twenty-five year perspective on establishment. Proceedings of the North American Prairie Conference 12: 169–178. University of Northern Iowa, Cedar Falls.

CHAPTER 7

ASSISTANCE PROGRAMS

Whether a basin is used for agriculture or restored to a wetland is often a matter of economics. Current programs, such as the Conservation Reserve Program (CRP) and Reinvest-in-Minnesota (RIM), have made it feasible for thousands of landowners in the region to restore wetlands. The relative ease of restoring wetlands in the prairie pothole region and the financial incentives have resulted in a level of landowner interest unsurpassed in the United States. Landowners can be compensated for long-term land retirement, and they can obtain technical assistance for planning and constructing a restoration from state and federal agencies. Additional support for seeding and construction costs may be available from conservation organizations such as Ducks Unlimited or local sportsmen's groups.

Landowners seeking assistance in restoring wetlands are generally expected to remove the basin and some surrounding area from agricultural use, including crop production, haying, and grazing. Land retirement options include long-term leases, permanent easements, and purchase agreements. The CRP initiated by the 1985 Farm Bill and administered by the USDA is a 10-year lease program. Permanent easements and 30-year lease arrangements were added to the Farm Bill in 1990 as part of the Wetland Reserve Program (WRP). In Minnesota, funding for permanent easements has been available since 1986 as part of the Reinvest-in-Minnesota program administered by the State of Minnesota Board of Water and Soil Resources. Additional opportunities for easements and purchase agreements may be available through the Prairie

141

Pothole Joint Venture. Private and governmental wetland interests in each state pool resources to increase waterfowl habitat and assist with the protection of high-priority sites through Prairie Pothole Joint Ventures.

Although lease or purchase agreements have traditionally been very popular ways to dedicate land for wetlands, permanent conservation easements are becoming increasingly common. Ten-year leases give landowners chances to restore wetlands even if they are unwilling to make a permanent commitment of land. Consequently, leases are an effective way to introduce large numbers of landowners to wetland restorations. Leases are less attractive for large, more complex projects because the costs of restoration are too high to justify only a short-term restoration. Very large marshes or wetland complexes are often high priorities for acquisition and restoration as Prairie Pothole Joint Venture projects and other major wetland conservation efforts.

Easements occupy somewhat of a "middle ground." They are more costly for wetland conservation interests to obtain than a lease but are cheaper than purchase. They require a permanent commitment from the landowner, but only for certain activities. Easements involve the transfer of certain rights relating to the use of the property. These land management prescriptions are specific and meant to ensure the quality and values of the site will be maintained. For instance, a conservation easement for a waterfowl nesting site will likely restrict conflicting uses such as haying. The value of the easement is decided by the nature of the restrictions. Besides describing restricted and allowed uses on the property, the terms of an easement include the length of the easement and provisions for monitoring the condition of the site. Landowners are not required to provide access to the public but will be required to provide access for the recipient of the easement.

CONSERVATION RESERVE PROGRAM

The CRP enrolls tracts that are classified as highly erodible or farmed wetlands by the Soil Conservation Service (SCS). Landowners bid to retire eligible property from crop production for 10 years with the Agricultural Stabilization and Conservation Service. The maximum acceptable CRP bid in an area is the conventional price for marginal-land cropping rights. If the bid is accepted, the landowner may choose to enroll in CRP by signing a binding agreement to receive an annual rental payment at the bid price, to develop and carry out a conservation plan for the property with consultation and approval by SCS, and to receive

50% federal cost sharing for establishing vegetative cover.

Since the CRP includes a variety of possible soil conservation measures, arranging for a wetland restoration is by a separate agreement during conservation planning. Another federal agency, the U.S. Fish and Wildlife Service can work with CRP-enrolled landowners to add special wetland restoration agreements to existing CRP leases. Federal or state wildlife biologists work with landowners to select potential restoration sites on CRP-enrolled land. Wildlife biologists then plan the project and often arrange the necessary construction work, and often can arrange for some financial assistance for the landowner's portion of construction costs. Some South Dakota wetland restoration programs give landowners the option of doing the construction and receiving payment.

WETLAND RESERVE PROGRAM

As with CRP, landowners bid to retire eligible land with their county Agricultural Stabilization and Conservation Service office. Eligible lands for the WRP are those classified as "prior converted wetlands" or "farmed wetlands" by the SCS. An upland buffer of no more than twice the wetland acreage may be enrolled, as long as the average width of adjacent land does not exceed 100 feet. The WRP allows for either 30-year leases or permanent easements. Priority will be given to farmers who agree to restore wetlands permanently. Landowners indicate their intentions to participate in the WRP with the Agricultural Stabilization and Conservation Service during an open enrollment period. Within 60 days, the SCS and other natural resource agencies work with the farmer to develop a restoration plan. The landowner can then choose to make a bid based on the plan, which outlines the terms of the easement. If the bid is accepted, the government will obtain a conservation easement from the landowner. The landowner can choose to receive payment for the easement in a single payment or payments spread over 10 years. The government will pay 75% of the cost of restoring the wetland. State agencies and private organizations may provide assistance for the remaining funds. As is typical with conservation easements, the contract will restrict some land uses but allow others; the landowner controls access to the land and does not have to open the property to public use. Unlike some easements, WRP land is subject to property taxes although these should be minimal because the land would no longer have agricultural value.

Lands next to the easement area on the same property cannot be managed in a way that will degrade the easement area; uses that could result in excessive sediment or agricultural chemical inputs will be restricted as will those that alter the flow of water into the easement area.

REINVEST IN MINNESOTA

RIM is a state program that has several wildlife habitat improvement goals, including wetland restorations. State of Minnesota Board of Water and Soil Resources staff plan the restoration and are responsible for construction and maintenance of any structures including all costs. Restorations are completed after the land has been retired with a conservation easement. Landowners work locally with county staff of Soil and Water Conservation Districts to initiate projects (during an annual enrollment period) and to document eligibility. Unlike the CRP and WRP, the value of the easement is determined by an appraisal rather than by bidding. The State Board of Water and Soil Resources receives all appraisals for the state and decides which projects it can fund.

Eligible lands must meet all of the following criteria: (1) is marginal farmland or next to marginal farmland, (2) was in agricultural crop production for 2 years from 1981 to 1985, (3) is owned by applicant for at least three years, (4) is at least 5 acres (minimum wetland restoration size for RIM is 1 acre with up to 4 acres of upland for each acre of wetland restored), (5) is not currently in another land retirement program.

Qualified, interested landowners obtain a single cash payment in exchange for a permanent wetland easement. As with the WRP, landowners are responsible for taxes, which should be minimal after land valuation changes because of use restrictions. After the easement has been acquired by the state and the land is part of the RIM reserve, the State Board of Water and Soil Resources develops detailed restoration plans and arranges necessary construction work and planting to permanent vegetative cover.

OTHER PROGRAMS

Most restorations within the region in the near future will likely be conducted under one of the three programs described here. However, other opportunities may develop if the popularity of private land restoration continues. There are currently private land assistance opportunities in each state, often through state Departments of Natural Resources.

Contact local agency staff for more information. For example, Iowa enacted a program in 1990 to protect wetlands. The Iowa Department of Natural Resources designates any wetland greater than 2 acres a protected wetland. Landowners of all protected wetlands must have permission to drain them but are eligible for a property tax credit on existing wetlands. The landowner applies for the tax exemption, which is issued as a certificate by the Department of Natural Resources. The certificate is submitted to the county treasurer who submits a claim to the department for reimbursement. Wetlands restored under an agreement with a local, state, or federal agency or private conservation group may be considered for tax exemption.

Some conservation opportunities are associated with the Farm Loan Program of the Farmers Home Administration. The Farm Debt Restructure and Conservation Set Aside Program allows the USDA to grant partial debt relief to the borrower for conservation easements of greater than 50 years duration.

WHO TO CONTACT

Federal land retirement programs, such as CRP and WRP, are administered through the USDA. District Conservationists at any county SCS office can provide information on these programs.

For assistance in wetland restoration planning and construction or information on Prairie Pothole Joint Venture projects, you can contact either state or federal wildlife offices. U.S. Fish and Wildlife Service Wetland Management Offices (Figure 7.1) and Department of Natural Resources Wildlife Offices (Figure 7.2) provide services for several counties.

Contact staff at county Soil and Water Conservation District Offices in Minnesota for information on RIM programs. They can also provide information on local sportsmen's groups active in wetland conservation in your area. In Iowa, county conservation boards are a good source of information on local wetland conservation activities.

The following private organizations are involved in various aspects of wetland protection, restoration, and management. They often organize funding and logistical support for wetland restorations: Ducks Unlimited, Pheasants Forever, Iowa Prairie Network, Iowa Natural Heritage Foundation, and The Nature Conservancy.

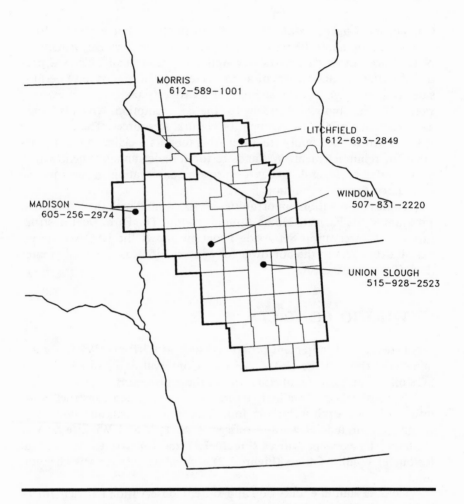

Figure 7.1. Wetland Management Offices of the U.S. Fish and Wildlife Service.

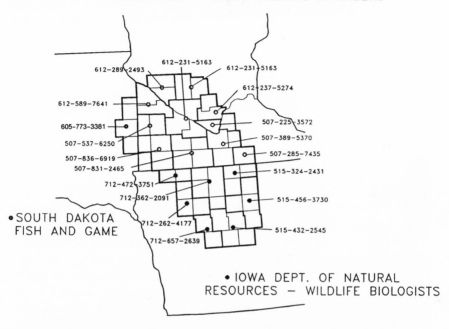

○ MINNESOTA DEPT. NATURAL
RESOURCES – AREA WILDLIFE MANAGERS

612–231–5163

612–289–2493

612–231–5163

612–237–5274

612–589–7641

605–773–3381

507–225–3572

507–537–6250

507–389–5370

507–836–6919

507–285–7435

507–831–2465

515–324–2431

712–472–3751

712–362–2091

515–456–3730

● SOUTH DAKOTA
FISH AND GAME

712–262–4177

515–432–2545

712–657–2639

● IOWA DEPT. OF NATURAL
RESOURCES – WILDLIFE BIOLOGISTS

Figure 7.2. Wildlife regions of state Departments of Natural Resources.

REFERENCES

Dunkle, F. and B. Misso. 1988. Farm bill related wetland protection and restoration opportunities. National Wetlands Newsletter, Jan–Feb: 3–5.

Hamilton, N.D. 1988. State initiatives to supplement the conservation reserve program. Report of the Legislative Extended Assistance Group, University of Iowa, Iowa City. 40 pp.

Steimel, D. 1992. Farmers drawn to wetlands offer. Des Moines Register, June 21, 1992. pp. 1B, 8B.

Ward, J.R. and F.K. Benfield. 1989. Conservation easements: prospects for sustainable agriculture. Virginia Environmental Law Journal 8: 271–292.

Wenzel, T. 1992. *Minnesota Wetlands Restoration Guide.* State of Minnesota Board of Water and Soil Resources. St. Paul, Minnesota.

SUMMARY

European settlers gladly traded wetlands and their plant and animal populations for more farmland and better roads, and waterfowl production yielded to agricultural production. We now realize that our forebears went too far and that too many wetlands were drained. We lost not only a significant part of our natural heritage but also the environmental, economic, and recreational benefits of these wetlands. Recreating the prairie pothole landscape as it existed 100 years ago, however, is not possible or desirable. A new balance between agricultural and conservation interests is developing, and this will result in the restoration of some of our lost natural heritage, including our wetlands. The new agricultural landscapes that are evolving hopefully will contain a mix of agriculture and natural systems. They will be a mosaic of crop fields, hay fields, pastures, prairies, and wetlands. Those parts of the landscape most suitable for crop production and the raising of livestock will be used for these purposes, but those not suitable for agriculture will be restored to native vegetation. This will reduce the environmental problems of agriculture and improve the quality of life of all those who live in the region.

If wetland restoration projects consist only of site selection and construction, our restored wetlands may never come to resemble natural wetlands in either composition or function. Breaking tile lines and plugging ditches are not necessarily all that are needed. Natural wetlands provide us with multiple benefits: recreational, aesthetic, historic, hydrologic, environmental, educational, etc. If the success of a wetland restoration is judged only by whether it contains any water or not or whether or not a duck has ever been seen on it, we will become satisfied with only partially restored wetlands at best. For example, a restored semi-permanent wetland without a sedge meadow, even if it is used by migrating ducks, is incomplete. Sedge meadows are full of small treasures, such as lilies, orchids, rails, and jumping mice, and are nesting sites for waterfowl. Only a restored semi-permanent wetland with a sedge meadow is a truly successful restoration because it is comparable in composition and function to a natural one. The challenge for restorationists is to apply our considerable scientific knowledge of wetland ecology to the selection of restoration sites, to designing restorations, to evaluating restorations, and to managing restorations to ensure that our restored wetlands are truly comparable to natural wetlands.

All restoration planning should be done within the context of a wetland complex to ensure that a mix of semi-permanent, seasonal, and temporary wetlands are restored in an area. Available funds would be better spent on restoring complexes than randomly selected isolated wetlands. Wetland complexes with a mosaic of temporary, seasonal, and semi-permanent wetlands will provide better habitat for wildlife as well as fulfill other important functions. The key to restoring wetland complexes is, obviously, cooperation among the agencies and groups involved in planning the restorations and cooperation among neighboring landowners. Viewing each project as part of a complex should reduce the temptation to restore every wetland as a semi-permanent wetland.

All restoration projects should have four phases: site selection, design and construction, evaluation, and management. Each phase has its own guidelines that need to be considered to maximize the probability that a project will result in a successful restoration. The restoration of wetlands in the prairie pothole region is a relatively recent phenomenon. We have only limited postconstruction experience on which to base these guidelines. Although it is likely that better or more complete guidelines can and will be developed, we offer these guidelines as a first approximation of more definitive guidelines that will evolve with time. We think that most of them will stand the test of time, however, because they are simply common sense and many have already been adopted.

SITE SELECTION

General site selection criteria can be either historic (to restore the wetland formerly present) or functional (to restore a wetland for a specific purpose, e.g., to improve water quality). Specific site selection criteria for projects will depend on the general criterion being used. In all cases, sites need to be located, their restoration feasibility determined, and required state and federal permits obtained. A base map must be made of all basins to be restored to determine the size of the basin, size of the watershed, locations and size of drainage tiles and/or ditches, and locations of property lines, roads, railroad lines, etc. A topographic survey is essential for basins with or adjacent to roads and property lines. Some important site selection guidelines are

1. Restoration sites should have hydric soils. This will ensure the areas once held water and may do so again after tiles are broken or ditches are plugged.
2. Potential restoration sites can be identified by locating hydric soils on county soils maps.
3. Basins with hydric soils that formed under groundwater discharge conditions (Histosols or peats) may not reflood because of a regional drop in groundwater levels.
4. Basins with hydric soils that formed under groundwater recharge conditions (Albolls) could become sites of groundwater contamination for restored wetlands in watersheds primarily in row crops.
5. Entire basins should be restored rather than parts of a larger basin.
6. Temporary and seasonal wetlands should have at least 2.5 acres of watershed for each acre of basin with a minimum 4 acres of watershed.
7. Semi-permanent wetlands should have at least 4 acres of watershed for each acre of basin with a minimum of 17 acres of watershed.
8. Restored wetlands will need to cover 0.5% of a watershed to affect flood-flows. Restoration of stream channels with fringe vegetation will do more to reduce flood-flows than restoring only a few small wetlands in a watershed.
9. Sites that were poorly drained, drained by ditches, drained by tile for less than 20 years, or used as pastures have the best revegetation potential.
10. Different groups of animals have specific habitat requirements that must be considered if a wetland is to be restored primarily as wildlife habitat. One group of animals should be chosen as the primary target group.

RESTORATION DESIGN

In the design of a restoration, dikes and water control structures must be provided that will control the location and extent of flooding;

dikes and water control structures must be properly sized and constructed; and any features required to optimize the wetland for a specific function need to be included. The following guidelines should be considered, when appropriate, in designing a restoration:

1. For tile-drained basins, all tile lines passing through the site must be broken and at least 50 feet removed or replaced with impermeable tile.
2. For ditch-drained basins, ditch plugs should be constructed from impervious material to minimize seepage, and ditches within the basin should be filled to grade.
3. Dikes should be constructed in accordance with established engineering guidelines.
4. Semi-permanent and permanent wetlands should have water control structures that can be used to drain the wetland.
5. A site that could have a good wetland seed bank should not be excavated.
6. Removing the surface layer of a basin during construction may be reasonable if the site received excessive sediments from severely eroding cropland. Do not excavate into the subsoil, particularly if the subsoil is sand or gravel.
7. Projects in watersheds with significant soil erosion problems should be avoided because restored wetlands in such watersheds will have only a short life span.
8. There should be a grassland buffer around all restored wetlands to prevent sediment from entering the restored wetland and to provide nesting habitat for upland breeding waterfowl.
9. Inlets and outlets in wetlands being restored to improve water quality should be as far apart as possible.
10. Nesting islands should be designed so that predators cannot get access to them.

EVALUATIONS

Regular visits to restored wetlands are important for two reasons. One, they enable any problems with a restoration to be quickly detected and appropriate action taken to correct them. Studies have shown that the probability of success of a wetland restoration and creation project increases significantly if there is post-construction monitoring. Two, they provide essential information about the development of restored wetlands. Very little is known about the revegetation and recolonization by animals of restored wetlands. Such information is needed to refine and improve site selection and design criteria. Even a brief and superficial annual visit that records the condition of dikes and water control structures is better than no visit. Simple surveys of the plant species present and periodic measurements of water depth are recommended annually for the first five years and every two or three years after that until there is little or no change in the composition of the vegetation. As part of an evaluation, the following should be considered:

1. Staff gauges should be installed in a deep portion of the basin and water depths recorded regularly.
2. Lists of all plant species in the basin should be made annually for five years after restoration and every two or three years after that until there is no further change in the species present.
3. The status of the vegetation in a restored marsh can be determined by comparing its species lists with those from natural marshes in Table 5.2.
4. The success of restored wetlands as wildlife habitat can be determined by doing a census of the wetlands for target species or specific groups of animals.
5. Evaluations can be used to identify structural, revegetation, or other problems with restoration that need to be corrected.

MANAGEMENT

Problems with restored wetlands are not unusual. They range from dike failure to clogged outlets to revegetation problems. A basic management plan should be developed for each restored wetland. At a minimum, it should include regular inspections of dikes and water control structures. Potential revegetation problems, including both the failure of certain groups of species to become established and the establishment of unwanted species, should be considered and contingency plans developed to deal with them. The following should be considered in developing management plans:

1. Dense litter from planted cover crops, such as switchgrass, may prevent the establishment of wetland species in newly restored basins. Burning the litter before the site is flooded will improve conditions for seed germination.
2. Turbidity due to algal blooms or fish can cause problems for the establishment and survival of submersed aquatics.

3. Restored wetlands should not be stocked with fish.
4. All inlets and outlets should be screened to prevent carp from entering restored wetlands.
5. If carp become a problem, they should be controlled with drawdowns, not with chemicals.
6. There are a number of weedy species, e.g., *Lythrum salicaria,* that should be eradicated from a restored wetland as soon as they are detected.
7. Restored wetlands with rare plants or fen plants should be managed carefully to preserve these species. Appropriate state and federal conservation organizations should be notified of their presence.
8. When required or desired, plant species can be established in five ways: (a) donor seed banks, (b) inoculation with donor soil, (c) seeding, (d) spreading wild hay, and (e) transplanting seedlings or adults.
9. Soil conservation practices should be encouraged to reduce the amount of sediment reaching the basin.
10. Limiting pesticide loading will help invertebrate production on which amphibians and some insectivorous mammals and many birds rely.

At least in the short term, the momentum generated by past achievements will make it feasible to continue to restore prairie potholes because funding for these restorations will continue. In the long term, however, funding for restoration programs will come increasingly to depend on the demonstrable success of past restorations. Restored wetlands must prove to be reasonable facsimiles of their natural counterparts or else there is little point in continuing to restore them. Restored wetlands will resemble natural wetlands only if they are thoughtfully and carefully planned, restored, and managed. They will require continued stewardship long after construction. Much information is currently available that can be used to improve the selection of restoration sites and the design of restorations, but it is often not fully used. More information, however, is needed on many aspects of the ecology of restored wetlands, particularly on sedge meadow revegetation, the fate of pesticides, contaminant effects on invertebrate and wildlife production, waterfowl predation, sediment impacts, groundwater inputs and outputs, and flood-flow alteration, to improve our understanding of how best to restore them.

PLANTS OF THE SOUTHERN PRAIRIE POTHOLE REGION

THIS SECTION includes a plant list for the southern prairie pothole region organized by family (A1), a rare plant list with county distributions (A2), and a key to the plants of the region (A3). This list was compiled from field observations, published plant lists, and reviews of field guides (listed in the following). Plant names generally follow *Flora of the Great Plains* (1986). The following people have contributed and/or reviewed information on the lists: Robert Kaul (University of Nebraska), Gary Larson (South Dakota State University), Deborah Lewis (Iowa State University), and Welby Smith (Minnesota Department of Natural Resources). William Norris (Iowa State University) reviewed the key.

Abbreviations Used in Appendix A Plant Lists

Zones (for explanation of each, see Chapter 2):

WP	Wet Prairie	FN	Fen
SM	Sedge Meadow	MF	Mudflat
SEM	Shallow Emergent Marsh	OW	Open Water
DEM	Deep Emergent Marsh	WO	Wooded Area

Groups (for explanation of each, see Chapter 5):

WP	Wet Prairie plants	FA	Floating Annuals
SM	Sedge Meadow plants	MF	Mudflat Annuals

155

SE	Shallow Emergent Marsh plants	WO	Woody Plants
DE	Deep Emergent Marsh plants	FN	Fen Plants
SA	Submersed Aquatics	FP	Floating Perennials

Families:

ACE	Maples	LTB	Bladderworts
ALS	Water Plantains	LYT	Loosestrifes
AMA	Pigweeds	MLV	Mallows
API	Parsleys	MNY	Buckbeans
APO	Dogbanes	NAJ	Naiads
ARA	Arums	NYM	Water Lilies
ASC	Milkweed	ONA	Evening Primroses
AST	Asters	ORC	Orchids
BRA	Mustards	PLG	Smartweeds
CAL	Water Starwort	PLM	Phloxes
CER	Hornworts	PLP	Ferns
CHN	Goosefoots	PNT	Pickerelweeds
CLU	St. Johnswort	POA	Grasses
CMP	Bellflowers	POT	Pondweeds
CRA	Stonecrops	PRM	Primroses
CRY	Pinks	RAN	Buttercups
CYP	Sedges	RIC	Liverworts
ELA	Waterworts	ROS	Roses
EQU	Horsetails	RUB	Madders
FAB	Beans	RUP	Ditchgrass
GEN	Gentians	SAL	Willows
HAL	Water Milfoils	SAX	Saxifrages
HIP	Mare's Tails	SCR	Figworts
HYD	Frog-bits	SPG	Burreeds
IRD	Irises	TYP	Cattails
JCG	Arrowgrasses	URT	Nettles
JUN	Rushes	VIO	Violets
LAM	Mints	VRB	Vervains
LIL	Lilies	ZAN	Horned Pondweeds
LMN	Duckweeds		

Abundance:

C   Common: Widespread throughout the region.
I   Infrequent: Found sporadically in the region.
R   Rare: Found in few locations within the region.

State Status:

E	Endangered		PT	Proposed Threatened
PE	Proposed Endangered		SC	Special Concern
T	Threatened		X	Extirpated

County Abbreviations:

Iowa (IA) Counties		Minnesota (MN) Counties		South Dakota (SD) Counties	
Boon	Boone	Blue	Blue Earth	Broo	Brookings
Buen	Buena Vista	Brow	Brown	Deue	Deuel
Calh	Calhoun	Chip	Chippewa	Mood	Moody
Carr	Carroll	Cott	Cottonwood		
Cerr	Cerro Gordo	Free	Freeborn		
Clay	Clay	Jack	Jackson		
Dick	Dickinson	Kand	Kandiyohi		
Emme	Emmet	LacQ	Lac Qui Parle		
Fran	Franklin	LeSe	LeSuer		
Gree	Greene	Linc	Lincoln		
Guth	Guthrie	Lyon	Lyon		
Hami	Hamilton	McLe	McLeod		
Hanc	Hancock	Mart	Martin		
Hard	Hardin	Meek	Meeker		
Humb	Humboldt	Murr	Murray		
Koss	Kossuth	Nico	Nicollet		
Osce	Osceola	Nobl	Nobles		
Palo	Palo Alto	Pipe	Pipestone		
Poca	Pocahontas	Redw	Redwood		
Sac	Sac	Renv	Renville		
Stor	Story	Sibl	Sibley		
Webs	Webster	Stea	Stearns		
Winn	Winnebago	Swif	Swift		
Wort	Worth	Wase	Waseca		
Wrig	Wright	Yell	Yellow Medicine		

Field Guides and References for Plant Identification

Fassett, N.C. 1957. *A Manual of Aquatic Plants.* University of Wisconsin Press, Madison.

Great Plains Flora Association. 1986. *Flora of the Great Plains.* University Press of Kansas, Lawrence. 1392 pp.

Pohl, R.W. 1975. *Keys to Iowa Vascular Plants.* Kendall/Hunt Publishing Company, Dubuque,Iowa. 198 pp.

van Bruggen, T. 1976. *Vascular Plants of South Dakota.* Iowa State University Press, Ames.

A1. PLANTS OF THE SOUTHERN PRAIRIE POTHOLE REGION

TABLE A1.1

ZONE	GROUP	SPECIES	FAM	COMMON NAME	SYNONYM	A
WO	WO	Acer negundo	ACE	Box elder		C
MF SEM	SE	Alisma gramineum J.G. Gmel.	ALS	Water plantain		R
SEM	SE	Alisma triviale Pursh	ALS	Water plantain	A. subcordatum / A. plantago-aquatica	C
MF SEM	SE	Sagittaria calycina Engelm.	ALS	Arrowhead	Lophotocarpus calycinus	I
MF SEM	SE	Sagittaria cuneata Sheldon	ALS	Arrowhead		C
MF SEM	SE	Sagittaria engelmanniana ssp. brevirostra	ALS	Arrowhead	S. brevirostra	R
MF SEM	SE	Sagittaria graminea Michx.	ALS	Arrowhead		R
MF SEM	SE	Sagittaria latifolia Willd.	ALS	Wapato		C
SEM	SE	Sagittaria rigida Pursh	ALS	Arrowhead		R
MF	MF	Amaranthus rudis Sauer	AMA	Pigweed	A. tamariscinus Nutt.	C
MF	MF	Amaranthus tuberculatus (Moq.) Sauer	AMA	Water hemp	Acnida altissima	I
SM FN	SM FN	Berula pusilla (Nutt.) Fern.	API	Water parsnip	B. erecta var. incisum	R
SM SEM	SE	Cicuta bulbifera L.	API	Bulbous water hemlock		I
WP SM	SM	Cicuta maculata L.	API	Water hemlock		C
WP	WP	Heracleum sphondylium L.	API	Cow parsnip		C
SM SEM	SE	Sium suave Walt.	API	Water parsnip		C
WP	WP	Zizia aptera (A. Gray) Fern.	API	Meadow parsnip		I
WP	WP	Zizia aurea (L.) Koch	API	Golden alexander		C
MF WP SM	SM	Apocynum cannabinum L.	APO	Indian hemp		C
SEM	SE	Acorus calamus L.	ARA	Sweet flag		I
SEM	SE	Peltandra virginica (L.) Schott.	ARA	Arrow arum		R
WP SM	SM	Asclepias incarnata L.	ASC	Swamp milkweed		C
MF SM	SM	Aster brachyactis Blake	AST	Rayless aster	Brachyactis ciliata	R
MF SM	SM	Aster falcatus Lindl. ssp. commutatus	AST	Aster		R
SM FN	SM FN	Aster junciformis Rydb.	AST	Rush aster	A. borealis	R
WP SM	SM	Aster novae-angliae L.	AST	New England aster		C
WP SM	SM	Aster puniceus L.	AST	Swamp aster	A. lucidulus	I
MF WP	WP	Aster simplex (Willd.) Jones	AST	Panicled aster	A. lanceolatus subsp. simplex	I
WP SM FN	SM FN	Aster pubentior Cronq.	AST	Flat top aster	A. umbellatus	I
MF	MF	Bidens cernua L.	AST	Sticktight		I
MF SM	SM FN	Bidens comosa (Gray) Wieg.	AST	Beggar's ticks		C
SM	SM	Bidens connata Muhl. ex Willd.	AST	Beggar's ticks	B. tripartita	I

TABLE A1.1. (*Continued*)

ZONE	GROUP	SPECIES	FAM	COMMON NAME	SYNONYM	A
SM	SM	*Bidens coronata (L.) Britt.*	AST	Tickseed sunflower	*B. trichosperma*	I
MF SM	SM	*Bidens frondosa L.*	AST	Beggar's ticks		C
MF	MF	*Bidens vulgata Greene*	AST	Beggar's Ticks		C
WP SM	SM	*Boltonia asteroides (L.) L'Her.*	AST	False aster		I
WP SM	SM	*Cirsium arvense (L.)Scop.*	AST	Canada thistle		C
WP	WP	*Cirsium discolor (Muhl. ex Willd.)*	AST	Field thistle		C
WP	WP	*Cirsium flodmanii (Rydb.) Arthur*	AST	Flodman's thistle		C
WP SM FN	SM	*Cirsium muticum Michx.*	AST	Swamp thistle		R
MF SM	SM	*Erechtites hieracifolia (L.) Raf.*	AST	Fireweed		I
MF WP	WP	*Erigeron philadelphicus L.*	AST	Fleabane		C
SM FN	SM FN	*Eupatorium maculatum L.*	AST	Joe pye weed		I
SM FN	SM FN	*Eupatorium perfoliatum L.*	AST	Boneset		C
WP SM	SM	*Helenium autumnale L.*	AST	Sneezeweed		C
WP SM	SM	*Helianthus grosseserratus Martens*	AST	Sawtooth sunflower		C
WP SM	SM	*Helianthus nuttallii ssp. rydbergii (Britt.) Long*	AST	Nuttall's sunflower		I
WP	WP	*Helianthus strumosus L.*	AST	Sunflower		I
WP	WP	*Helianthus tuberosus L.*	AST	Jerusalem artichoke		I
WP	WP	*Liatris lancifolia (Greene) Kitt.*	AST	Blazing star		I
WP	WP	*Liatris ligulistylis (Nels.) Schum.*	AST	Blazing star		C
WP	WP	*Liatris pycnostachya Michx.*	AST	Prairie gayfeather		C
OW	SA	*Megalodonta beckii (Torr.) Greene*	AST	Water marigold	*Bidens beckii*	R
WP	WP	*Ratibida pinnata (Vent.) Barnh.*	AST	Gray coneflower		C
WP	WP	*Rudbeckia hirta L.*	AST	Black-eyed susan		C
WP SM	SM	*Senecio aureus L.*	AST	Golden ragwort	*S. pseudaurens*	I
MF	MF	*Senecio congestus (R.Br.)DC.*	AST	Marsh fleabane		R
WP	WP	*Senecio integerrimus Nutt.*	AST	Ragwort		C
WP	WP	*Senecio pauperculus Michx.*	AST	Prairie ragwort		C
WP	WP	*Silphium laciniatum L.*	AST	Compass plant		C
WP	WP	*Silphium perfoliatum L.*	AST	Cup plant		C
WP	WP	*Solidago gigantea Ait.*	AST	Late goldenrod		C
WP FN	WP FN	*Solidago graminifolia*	AST	Goldenrod	*Euthamia graminifolia*	I
WP SM FN	SM FN	*Solidago riddellii Frank*	AST	Riddell's goldenrod		R

TABLE A1.1. (Continued)

ZONE	GROUP	SPECIES	FAM	COMMON NAME	SYNONYM	A
WP SM	SM	*Vernonia fasciculata Michx.*	AST	Ironweed		I
MF	MF	*Xanthium strumarium L.*	AST	Sniny cocklebur		C
SM FN	SM FN	*Cardamine bulbosa BSP.*	BRA	Spring cress		I
SM	SM	*Cardamine pensylvanica Muhl. ex Willd.*	BRA	Bitter cress		C
MF SM SEM	MF	*Rorippa palustris (L.) Besser.*	BRA	Marsh cress	*R. islandica*	C
MF SEM	SA	*Callitriche heterophylla Pursh.*	CAL	Water starwort		R
MF SEM	SA	*Callitriche verna L.*	CAL	Water starwort	*C. palustris*	R
DEM OW	SA	*Ceratophyllum demersum L.*	CER	Coontail		C
MF	MF	*Chenopodium album L.*	CHN	Lambs-quarters		C
MF	MF	*Chenopodium glaucum L.*	CHN	Oak-leaved goosefoot		I
MF	MF	*Chenopodium rubrum L.*	CHN	Alkali blite		I
SM	SM	*Hypericum majus (Gray)Britt.*	CLU	St. Johnswort		I
SM	SM	*Campanula aparinoides Pursh*	CMP	Eastern marsh bellflower		R
FN	FN	*Lobelia kalmii L.*	CMP	Lobelia		C
WP SM	SM	*Lobelia siphilitica L.*	CMP	Great lobelia		C
WP	WP	*Lobelia spicata Lam.*	CMP	Spiked lobelia		C
MF SM	SM	*Cerastium nutans L.*	CRY	Prairie chickweed		C
SM FN	SM FN	*Stellaria crassifolia Ehrh.*	CRY	Fleshy stitchwort		C
SM FN	SM FN	*Stellaria longifolia Muhl. ex Willd.*	CRY	Long-leaved stitchwort		R
SM	SM	*Carex alopecoidea Tuckerm.*	CYP	Foxtail sedge		R
SM FN	SM FN	*Carex aquatilis Wahl.*	CYP	Water sedge		C
SM SEM	SE	*Carex atherodes Spreng.*	CYP	Awned sedge	*C. substricta*	R
WP SM	SM	*Carex aurea Nutt.*	CYP	Sedge		I
WP SM	SM	*Carex bebbii (Bailey) Fern.*	CYP	Bebb's sedge		I
WP SM	SM	*Carex bicknellii Britt.*	CYP	Bicknell's sedge		I
SM FN	SM FN	*Carex buxbaumii Wahl.*	CYP	Sedge		I
SM SEM	SM FN	*Carex comosa Boott*	CYP	Bristly sedge		I
WP SM	SM	*Carex crawei Dew.*	CYP	Sedge		I
WP SM	SM	*Carex cristatella Britt.*	CYP	Crested sedge		I
SM FN	SM FN	*Carex diandra Schrank.*	CYP	Sedge		R
WP	WP	*Carex douglasii Boott*	CYP	Sedge		I
SM	SM	*Carex emoryi Dewey*	CYP	Emory's sedge	*C. stricta var. elongata*	I

TABLE A1.1. (Continued)

ZONE	GROUP	SPECIES	FAM	COMMON NAME	SYNONYM	A
SM	SM	Carex granularis Muhl. ex Willd.	CYP	Sedge		I
WP SM	WP	Carex gravida Bailey	CYP	Heavy sedge		I
WP SM	SM	Carex hallii Olney	CYP	Hall's sedge		R
SM	SM	Carex haydenii Dewey	CYP	Sedge		I
SM FN	SM FN	Carex hystericina Muhl ex. Willd.	CYP	Porcupine sedge		C
WP SM FN	SM FN	Carex interior L. Bailey	CYP	Inland sedge		I
SEM	SE	Carex lacustris Willd.	CYP	Lake sedge		C
WP SM SEM	SEM SE	Carex laeviconica Dewey	CYP	Smooth-sheathed awned sedge	C. trichocarpa var. deweyi	I
WP SM	SM	Carex lanuginosa Michx.	CYP	Woolly sedge		C
WP SM	SM	Carex praegracilis Boot.	CYP	Clustered field sedge		I
SM FN	SM FN	Carex prairsa Dewey	CYP	Prairie sedge	C. prairea	I
WP SM	SM	Carex projecta Mack.	CYP	Sedge		R
SM	SM	Carex retrorsa Schw.	CYP	Retrorse sedge		I
SM SEM FN	SE	Carex rostrata Stokes	CYP	Beaked sedge		R
SM	SM	Carex sartwellii Dewey	CYP	Sartwell's sedge		C
SM FN	SM FN	Carex scoparia Schkuhr ex. Willd.	CYP	Pointed broom sedge		R
FN	SM FN	Carex sterilis Willd.	CYP	Sterile sedge		C
SM	SM	Carex stipata Muhl ex Willd.	CYP	Awn-fruited sedge		C
SM FN	SM FN	Carex stricta Lam.	CYP	Tussock sedge		C
MF SM	SM	Carex suberecta (Olney) Britt.	CYP	Prairie straw sedge		I
SM FN	SM	Carex synchnocephala Carey	CYP	Dense long-beaked sedge		I
SM	SM	Carex tetanica Schkuhr.	CYP	Wood's sedge		I
SM	SM	Carex tribuloides Wahl.	CYP	Blunt broom sedge		R
SM	SM	Carex trichocarpa Schkuhr.	CYP	Hairy-fruited sedge		I
SM SEM	SE	Carex vesicaria L.	CYP	Inflated sedge		I
SM FN	SM FN	Carex viridula Michx.	CYP	Necklace sedge		I
MF WP SM	SM	Carex vulpinoidea Michx.	CYP	Fox sedge		C
MF	MF	Cyperus acuminatus Torr. & Hook.	CYP	Nutgrass		I
MF	MF	Cyperus aristatus Rottb.	CYP	Awned cyperus		C
MF	MF	Cyperus diandrus Torr.	CYP	Low cyperus		I
MF	MF	Cyperus engelmannii Steud.	CYP	Engelmann's cyperus		I
MF	MF	Cyperus erythrorhizos Muhl.	CYP	Red-rooted cyperus		I

TABLE A1.1. (*Continued*)

ZONE	GROUP	SPECIES	FAM	COMMON NAME	SYNONYM	A
MF	MF	*Cyperus esculentus L.*	CYP	Yellow nutgrass		C
MF	MF	*Cyperus odoratus L.*	CYP	Fragrant cyperus		C
MF FN	FN	*Cyperus rivularis Kunth.*	CYP	Shining sedge		I
MF SM	SM	*Cyperus strigosus L.*	CYP	Nutgrass		C
FN	FN	*Dulichium arundinaceum (L.) Britt.*	CYP	Three-way sedge		R
MF	MF	*Eleocharis acicularis (L.) R.&S.*	CYP	Needle spike rush		C
WP SM	SM	*Eleocharis compressa Sulliv.*	CYP	Flat-stemmed spike rush		I
MF SM	SM	*Eleocharis erythropoda Steud.*	CYP	Spike rush	*E. calva*	C
SM SEM	SE	*Eleocharis macrostachya Britt.*	CYP	Pale spike rush		C
MF	MF	*Eleocharis obtusa (Willd.)Schultes*	CYP	Blunt spike rush		C
MF	MF	*Eleocharis parvula (R. & S.) Link*	CYP	Small spike rush	*E. coloradensis*	R
FN	FN	*Eleocharis pauciflora (Lightf.) Lin.*	CYP	Spike rush		R
SM SEM	SE	*Eleocharis smallii Britt.*	CYP	Marsh spike rush	*E. palustris*	C
WP SM FN	SM FN	*Eleocharis wolfii Gray*	CYP	Spike rush		R
SM FN	SM FN	*Eriophorum polystachion L.*	CYP	Tall cotton grass	*E. angustifolium*	I
SM FN	SM FN	*Eriophorum gracile Koch.*	CYP	Slender cotton grass		R
FN	FN	*Rhynchospora capillacea Torr.*	CYP	Beak rush		R
SEM	DE	*Scirpus acutus (Muhl) Bigel.*	CYP	Hard-stemmed bulrush		C
WP SM	SM	*Scirpus atrovirens Willd.*	CYP	Green bulrush		C
SM	SM	*Scirpus cyperinus (L.) Kunth.*	CYP	Wool bulrush	*S. pallidum*	I
SEM	SE	*Scirpus fluviatilis (Torr.) Gray*	CYP	River bulrush		C
SEM	DE	*Scirpus heterochaetus Chase*	CYP	Slender bulrush		I
SM SEM	SE	*Scirpus maritimus L.*	CYP	Prairie bulrush	*S. paludosus*	I
SEM	SE	*Scirpus pungens Vahl.*	CYP	Three square bulrush		I
MF	MF	*Scirpus smithii*	CYP	Smith's sedge		R
SEM	DE	*Scirpus validus Vahl.*	CYP	Soft-stemmed bulrush		C
MF	MF	*Elatine triandra Schkuhr.*	ELA	Waterwort		R
WP SM MF	SM	*Equisetum arvense L.*	EQU	Field horsetail		C
SM SEM	SE	*Equisetum fluviatile L.*	EQU	Smooth scouring rush		I
WP SM	SM	*Equisetum hyemale L.*	EQU	Common scouring rush		C
SM	SM	*Equisetum pratense Ehrh.*	EQU	Meadow horsetail		C
WP SM	SM	*Amorpha fruticosa L.*	FAB	Indigo bush		C

TABLE A1.1. (Continued)

ZONE	GROUP	SPECIES	FAM	COMMON NAME	SYNONYM	A
WP	WP	Desmodium canadense (L.) DC.	FAB	Tick clover		C
WP SM	SM	Lathyrus palustris L.	FAB	Marsh vetchling		C
WP SM	SM	Gentiana andrewsii Griseb.	GEN	Bottle gentian		I
WP	WP	Gentiana puberulenta Pringle	GEN	Prairie gentian		I
SM FN	SM FN	Gentianopsis crinita Froel.	GEN	Fringed gentian	Gentiana crinita	R
SM FN	SM FN	Gentianopsis procera Holm.	GEN	Small fringed gentian	Gentiana procera	R
DEM	SA	Myriophyllum heterophyllum Michx.	HAL	Water milfoil		R
DEM OW	SA	Myriophyllum spicatum L.	HAL	Water milfoil	M. exalbescens	I
DEM OW	SA	Myriophyllum verticillatum L.	HAL	Water milfoil		R
MF SEM	SA	Hippuris vulgaris L.	HIP	Mare's tail		R
DEM OW	SA	Elodea canadensis Michx.	HYD	Waterweed		I
DEM OW	SA	Elodea nuttallii (Planch.) St. John	HYD	Waterweed		I
DEM OW	SA	Vallisneria americana Michx.	HYD	Wild celery		R
SM SEM	SE	Iris shrevei Sm.	IRD	Blue flag	I. versicolor, I. virginica	C
WP FN	WP FN	Triglochin maritimum L.	JCG	Large arrow grass		I
FN	FN	Triglochin palustre L.	JCG	Marsh arrow grass		R
FN	FN	Juncus alpinoarticulatus Chaix in Villars	JUN	Rush	J. alpinus	R
WP SM FN	SM FN	Juncus balticus Willd.	JUN	Bog rush		I
MF	MF	Juncus bufonis L.	JUN	Toad rush		R
MF SM	SM	Juncus dudleyi Wieg.	JUN	Dudley rush		C
MF SM	SM	Juncus interior Wieg.	JUN	Inland rush		I
WP	WP	Juncus nodosus L.	JUN	Knotted rush		C
MF SM	SM	Juncus torreyi Cov.	JUN	Torrey's rush		C
MF SM	SM	Lycopus americanus Muhl.	LAM	Water horehound		C
WP SM	SM	Lycopus asper Greene	LAM	Rough bugleweed		I
WP SM	SM	Lycopus uniflorus Michx.	LAM	Northern bugleweed		C
WP SM	SM	Mentha arvensis L.	LAM	Field mint		C
WP	WP	Physostegia virginiana (L.) Benth	LAM	False dragonhead		I
WP MF	WP	Pycnanthemum virginianum (L.) Dur. & Jack.	LAM	Mountain mint		C
SM	SM	Scutellaria galericulata L.	LAM	Common skullcap		C
SM	SM	Scutellaria lateriflora L.	LAM	Mad-dog skullcap		C
WP SM	SM	Stachys palustris L.	LAM	Woundwort		C

TABLE A1.1. (Continued)

ZONE	GROUP	SPECIES	FAM	COMMON NAME	SYNONYM	A
WP SM	SM	Teucrium canadense L.	LAM	American germander		C
WP	WP	Allium canadense L.	LIL	Wild garlic		C
WP	WP	Hypoxis hirsuta (L.) Cov.	LIL	Yellow star grass		C
WP	WP	Lilium michiganense Farw.	LIL	Michigan lily		I
WP	WP	Lilium philadelphicum var. andinum L.	LIL	Wood lily		I
WP	WP	Zigadenus elegans Pursh.	LIL	Death camas		I
SEM	FA	Lemna minor L.	LMN	Lesser duckweed		C
SEM DEM	FA	Lemna trisulca L.	LMN	Star duckweed		C
SEM	FA	Spirodela polyrrhiza (L.) Schleiden	LMN	Greater duckweed		C
SEM	FA	Wolffia columbiana Karst.	LMN	Watermeal		C
FN SEM	SA	Utricularia intermedia Hayne	LTB	Flat-leaved bladderwort		R
FN	FN	Utricularia minor L.	LTB	Small bladderwort		R
SEM DEM	SA	Utricularia vulgaris L.	LTB	Common bladderwort		C
MF	MF	Ammania coccinea Rottb.	LYT	Toothcup		I
MF	MF	Ammania robusta Heer. & Regel.	LYT	Toothcup		I
SM	SM	Lythrum alatum Pursh.	LYT	Common loosestrife		I
MF SM SEM	SE	Lythrum salicaria L.	LYT	Purple loosestrife		I
MF	MF	Rotala ramosior (L.) Koehne	LYT	Toothcup		R
MF	MF	Hibiscus militaris Cav.	MLV	Rose mallow		I
SM FN	SM FN	Menyanthes trifoliata L.	MNY	Bog bean		R
DEM OW	SA	Najas flexilis (Willd.) Rostk. & Schmidt	NAJ	Bushy pondweed		C
DEM OW	SA	Najas guadalupensis (Spreng.) Magnus.	NAJ	Bushy pondweed		I
DEM OW	SA	Najas marina L.	NAJ	Bushy pondweed		R
DEM OW	FP	Brasenia schreberi Gmelin	NYM	Water shield		R
DEM OW	FP	Nuphar luteum (L.) Sibth & Sm. ssp. variegatum (Engelm ex. Clinton) Beal	NYM	Yellow water lily		I
DEM OW	FP	Nymphaea tuberosa Paine	NYM	White water lily	N. odorata	I
SM	SM FN	Epilobium ciliatum Raf.	ONA	Willow herb		I
SM FN	SM FN	Epilobium coloratum Biehler	ONA	Cinnamon willow herb		I
SM FN	SM FN	Epilobium leptophyllum Ref.	ONA	Bog willow herb		I
SM SEM	SE	Ludwigia polycarpa Short & Peter	ONA	False loosestrife		I
WP SM FN	SM FN	Cypripedium candidum Muhl. ex. Willd.	ORC	White lady slipper		R
WP SM	SM	Cypripedium reginae Walt.	ORC	Showy lady slipper		R

TABLE A1.1. (Continued)

ZONE	GROUP	SPECIES	FAM	COMMON NAME	SYNONYM	A
SM FN	SM FN	Liparis loeselii (L.) Rich.	ORC	Green twayblade		R
FN	FN	Platanthera hyperborea (L.) R. Br.	ORC	Northern green orchid	Habenaria hyperborea	R
WP SM	SM	Platanthera praeclara Shev. & Bowl.	ORC	Western prairie fringed orchid	Habenaria leucophea	R
SM	SM	Spiranthes cernua (L.) Rich.	ORC	Ladies tresses		R
WP SM	SM	Spiranthes magnicamporum Sheviak	ORC	Great Plains ladies tresses		R
FN	FN	Spiranthes romanzoffiana Cham.	ORC	Hooded ladies tresses		R
WP	WP	Spiranthes vernalis Engelm. & Gray	ORC	Twisted ladies tresses		R
WP SM SEM SE	SE	Polygonum amphibium L.	PLG	Water smartweed		C
MF SM	SM	Polygonum hydropiper L.	PLG	Common smartweed		C
MF	SM	Polygonum lapathifolium L.	PLG	Nodding smartweed		C
MF	MF	Polygonum pensylvanicum L.	PLG	Pinkweed		C
MF	MF	Polygonum persicaria L.	PLG	Lady's thumb		C
SM MF	SM	Polygonum punctatum Ell.	PLG	Water smartweed		I
MF	MF	Polygonum ramosissimum Michx.	PLG	Bushy knotweed		I
SM MF	SM	Rumex altissimus Wood.	PLG	Pale dock		C
WP SM MF	SM	Rumex crispus L.	PLG	Curly dock		C
SM MF	MF	Rumex maritimus L. var. fueginus	PLG	Golden dock		C
SM MF	SM	Rumex mexicanus Meisn.	PLG	Willow-leaved dock		I
SM MF	SM	Rumex orbiculatus Gray	PLG	Dock		C
SM	SM	Rumex stenophyllus Ledeb.	PLG	Dock		I
WP SM MF	SM	Rumex verticillatus L.	PLG	Water dock		I
WP	WP	Phlox pilosa L.	PLM	Prairie phlox		C
WP SM	SM	Onoclea sensibilis L.	PLP	Sensitive fern		I
WP SM	SM	Thelypteris palustris Schott.	PLP	Marsh fern		I
MF OW	SA	Heteranthera dubia (Jacq.) Macm.	PNT	Water star grass	Zosterella dubia	I
MF	MF	Heteranthera limosa (Sw.) Willd.	PNT	Mud plantain		R
SEM	SE	Pontederia cordata L.	PNT	Pickerelweed		I
WP SM	WP	Agropyron repens (L.) Beauv.	POA	Quackgrass		C
MF	MF	Agrostis hyemalis (Walt.) B.S.P.	POA	Ticklegrass		C
MF	MF	Agrostis stolonifera L.	POA	Redtop		C
MF SEM	SE	Alopecurus aequalis Sobol.	POA	Marsh foxtail		I
WP	WP	Andropogon gerardii Vitman	POA	Big bluestem		C

TABLE A1.1. (*Continued*)

ZONE	GROUP	SPECIES	FAM	COMMON NAME	SYNONYM	A
SM SEM	SE	*Beckmannia syzigachne (Steud.) Fern.*	POA	American sloughgrass		I
SM FN	SM FN	*Bromus ciliatus L.*	POA	Fringed brome		I
WP SM	SM	*Calamagrostis canadensis Beauv.*	POA	Bluejoint		C
SM FN	SM FN	*Calamagrostis stricta (Timm.) Koel.*	POA	Northern reedgrass	*C. neglecta, C. inexpansa*	I
MF	MF	*Echinochloa crusgalli (L.) Beauv.*	POA	Barnyard grass		C
MF	MF	*Echinochloa muricata (Beauv.) Fern.*	POA	Barnyard grass		C
MF	MF	*Eragrostis hypnoides (Lam.) B.S.P.*	POA	Teal lovegrass		C
MF SEM	SE	*Glyceria borealis (Nash.) Batch.*	POA	Northern manna grass		I
SM SEM	SE	*Glyceria grandis S. Wats.*	POA	Tall manna grass	*G. maxima*	C
SM SEM	SE	*Glyceria striata (Lam.) Hitchc.*	POA	Fowl manna grass		C
WP SM	SM	*Hierochloe odorata (L.) Beauv.*	POA	Sweetgrass		I
MF	MF	*Hordeum pusillus L.*	POA	Foxtail barley		C
SM MF	SM	*Leersia oryzoides (L.) Swartz*	POA	Rice cut grass		C
MF	MF	*Leptochloa fascicularis (Lam.) A. Gray*	POA	Bearded sprangleop		I
SM MF FN	SM FN	*Muhlenbergia asperifolia Nees & Meyer*	POA	Scratchgrass		I
SM FN	SM FN	*Muhlenbergia glomerata (Willd.) Trin.*	POA	Muhly		I
WP SM FN	SM FN	*Muhlenbergia mexicana (L.) Trin.*	POA	Mexican muhly		I
WP SM	SM FN	*Muhlenbergia racemosa (Michx.) B.S.P.*	POA	Marsh muhly		I
MF	MF	*Panicum capillare L.*	POA	Common witchgrass		C
SM	SM	*Panicum gattingeri Nash.*	POA	Panicle grass		R
WP	WP	*Panicum virgatum L.*	POA	Switchgrass		C
SM SEM	SE	*Phalaris arundinacea L.*	POA	Reed canary grass		C
SEM DEM	DE	*Phragmites communis Trin.*	POA	Reed	*Phragmites australis*	C
WP SM	SM	*Poa palustris L.*	POA	Fowl meadow grass		I
WP	WP	*Poa pratensis L.*	POA	Kentucky blue grass		C
SM SEM	SE	*Scolochloa festucacea (Willd.) Link*	POA	White top		C
WP SM	SM	*Spartina pectinata Link*	POA	Cord grass		I
WP SM	SM	*Sphenopholis obtusata (Michx.) Scribn.*	POA	Prairie wedge grass		I
SEM	DE	*Zizania aquatica L. var. interior Fass.*	POA	Wild rice		R
OW	SA	*Potamogeton amplifolius Tuckerm.*	POT	Large-leaved pondweed		R
OW	SA	*Potamogeton berchtoldii Raf.*	POT	Baby pondweed		R
OW	SA	*Potamogeton epihydrus Raf.*	POT	Nuttall's pondweed	*P. pusillus var. tenuissimus*	R

TABLE A1.1. *(Continued)*

ZONE	GROUP	SPECIES	FAM	COMMON NAME	SYNONYM	A
OW	SA	*Potamogeton filiformis Pers.*	POT	Slender pondweed		R
DEM OW	SA	*Potamogeton foliosus Raf.*	POT	Leafy pondweed		C
OW	SA	*Potamogeton friesii Rupr.*	POT	Fries' pondweed		R
DEM OW	SA	*Potamogeton gramineus L.*	POT	Variable pondweed		I
OW	SA	*Potamogeton illinoensis Morong.*	POT	Illinois pondweed		I
OW	SA	*Potamogeton natans L.*	POT	Floating pondweed		I
DEM OW	SA	*Potamogeton nodosus Poir.*	POT	Long-lived pondweed	*P. americanus*	C
DEM OW	SA	*Potamogeton pectinatus L.*	POT	Sago pondweed		C
OW	SA	*Potamogeton praelongus Wulfen*	POT	White-stemmed pondweed		R
DEM OW	SA	*Potamogeton pusillus L.*	POT	Baby pondweed		C
DEM OW	SA	*Potamogeton richardsonii (Benn.) Rydb.*	POT	Richardson's pondweed		R
DEM OW	SA	*Potamogeton spirillus Tuckerm.*	POT	Pondweed		R
DEM OW	SA	*Potamogeton strictifolius Benn.*	POT	Narrowleaf pondweed		R
DEM OW	SA	*Potamogeton vaginatus Turcz.*	POT	Sheathed pondweed		R
OW	SA	*Potamogeton vaseyi Robbins*	POT	Vasey's pondweed		R
DEM OW	SA	*Potamogeton zosteriformis*	POT	Flat-stemmed pondweed		C
WP SM	SM	*Lysimachia ciliata L.*	PRM	Fringed loosestrife		C
WP SM	SM	*Lysimachia hybrida Michx.*	PRM	Loosestrife		C
WP SM	SM	*Lysimachia quadriflora Sims.*	PRM	Whorled loosestrife		C
WP SM	SM	*Lysimachia terrestris (L.) BSP*	PRM	Swamp loosestrife		I
SM SEM	SM	*Lysimachia thyrsiflora L.*	PRM	Tufted loosestrife		I
WP	WP	*Anemone canadensis L.*	RAN	Canada anemone		C
SM FN	SM FN	*Caltha palustris L.*	RAN	Marsh marigold		C
SEM DEM	SA	*Ranunculus circinatus Sibh. var. subrigidus (Drew) Benson*	RAN	White water crowfoot	*R. subrigidus*	R
SM MF	SM	*Ranunculus cymbalaria Pursh*	RAN	Seaside crowfoot		C
MF SEM	SA	*Ranunculus flabellaris Raf.*	RAN	Yellow water crowfoot		I
SEM	SA	*Ranunculus gmelini DC.*	RAN	Small yellow crowfoot		R
DEM	SA	*Ranunculus longirostris Godr.*	RAN	White water crowfoot		I
SM MF	SM	*Ranunculus pensylvanicus L.*	RAN	Bristly crowfoot		I
MF	MF	*Ranunculus sceleratus L.*	RAN	Cursed crowfoot		C
WP	WP	*Thalictrum dasycarpum Fisch. and Lall.*	RAN	Tall meadow rue		C
SEM	FA	*Riccia fluitans L.*	RIC	Slender riccia		C

TABLE A1.1. (*Continued*)

ZONE	GROUP	SPECIES	FAM	COMMON NAME	SYNONYM	A
SEM	FA	*Ricciocarpus natans (L.) Corda*	RIC	Purple-fringed riccia		C
WP	WP	*Geum canadense Jacq.*	ROS	White avens		C
WP SM	SM	*Geum laciniatum Murr.*	ROS	Rough avens		I
WP	WP	*Geum macrophyllum Willd.*	ROS	Large-leaved avens		R
WP	WP	*Geum triflorum Pursh*	ROS	Purple avens		I
WP FN	WP FN	*Potentilla anserina L.*	ROS	Silverweed		I
WP SM	SM	*Potentilla norvegica L.*	ROS	Norwegian cinquefoil		C
SM FN	SM FN	*Potentilla palustris (L.) Scop*	ROS	Marsh cinquefoil		R
MF	MF	*Potentilla paradoxa Nutt.*	ROS	Bushy cinquefoil		R
MF	MF	*Potentilla rivalis Nutt.*	ROS	Cinquefoil		I
WP SM	SM	*Spiraea alba Du Rois*	ROS	Meadow sweet		I
WP	WP	*Galium boreale L.*	RUB	Northern bedstraw		I
WP	WP	*Galium obtusum Bigel.*	RUB	Bluntleaf bedstraw		C
SM SEM	SM	*Galium trifidum L.*	RUB	Small bedstraw		R
DEM OW	SA	*Ruppia occidentalis S. Wats.*	RUP	Widgeon grass	*R. maritima*	R
WO	WO	*Populus deltoides Marsh.*	SAL	Cottonwood		C
WO	WO	*Salix amygdaloides Anders.*	SAL	Peach-leaved willow		C
WO	WO	*Salix bebbiana Sarg.*	SAL	Long-beaked willow		I
FN	WO FN	*Salix candida Fluegge ex Willd.*	SAL	Hoary willow		R
WO	WO	*Salix discolor Muhl.*	SAL	Pussy willow		C
WO SM	WO	*Salix exigua Nutt.*	SAL	Sandbar willow	*S. interior*	C
WO	WO	*Salix fragilis L.*	SAL	Crack willow		I
WO	WO	*Salix lucida Muhl.*	SAL	Shining willow		R
WO	WO	*Salix nigra Marsh.*	SAL	Black willow		C
SM FN	WO	*Salix pedicellaris Pursh.*	SAL	Bog willow		R
SM FN	WO FN	*Salix petiolaris Smith*	SAL	Slender willow		C
WO	WO	*Salix rigida Muhl.*	SAL	Meadow willow		C
WP SM	WO	*Salix serissima (Bailey) Fern.*	SAL	Autumn willow	*S. lutea, S. eriocephala*	R
SM FN	SM FN	*Parnassia glauca Raf.*	SAX	Grass of parnassus	*Parnassia caroliniana*	R
WP SM MF	SM	*Penthorum sedoides L.*	SAX	Ditch stone crop		C
WP SM	SM	*Agalinis tenuifolia (Vahl.) Raf.*	SCR	Gerardia	*Gerardia tenuifolia*	I
MF SEM	SE	*Bacopa rotundifolia (Michx.) Wettst.*	SCR	Water hyssop		R

TABLE A1.1. (Continued)

ZONE	GROUP	SPECIES	FAM	COMMON NAME	SYNONYM	A
SM	SM	Chelone glabra L.	SCR	White turtlehead		R
WP SM	MF	Gratiola neglecta Torr.	SCR	Hedge hyssop		I
MF	MF	Limosella aquatica L.	SCR	Mudwort		R
MF OW	MF	Lindernia dubia (L.) Pennell	SCR	False pimpernel		I
FN	FN	Mimulus glabratus HBK. var. fremontii (Benth.)	SCR	Yellow monkey flower		R
SM MF	SM	Mimulus ringens L.	SCR	Monkey flower		C
WP SM FN	SM FN	Pedicularis lanceolata Michx.	SCR	Swamp lousewort		I
WP	WP	Veronicastrum virginicum (L.) Farw.	SCR	Culver's root		C
SM MF	SM	Veronica anagallis-aquatica L.	SCR	Water speedwell		C
MF SEM	SE	Veronica catenata Penn.	SCR	Speedwell		I
SM MF	SM	Veronica peregrina L.	SCR	Purslane speedwell		C
SEM	SE	Sparganium americanum Nutt.	SPG	Nuttall's burreed		R
SEM	SE	Sparganium chlorocarpum Rydb.	SPG	Green-fruited burreed		I
SEM DEM	SE	Sparganium eurycarpum Engelm.	SPG	Giant burreed		C
SEM DEM	DE	Typha angustifolia L.	TYP	Narrow leaf cattail		C
SEM DEM	DE	Typha x glauca Godron	TYP	Hybrid cattail		C
SEM	SE	Typha latifolia L.	TYP	Broad leaf cattail		C
SM FN	SM FN	Pilea fontana (Lunell) Rydb.	URT	Clearweed		R
WP SM	SM	Urtica dioica L.	URT	Stinging nettle		C
WP SM FN	SM FN	Viola nephrophylla Greene	VIO	Bog violet		I
SM FN	SM FN	Phyla lanceolata (Michx.) Greene	VRB	Fog fruit	Lippia lanceolata	C
WP SM	SM	Verbena hastata L.	VRB	Blue vervain		C
DEM OW	SA	Zannichellia palustris L.	ZAN	Horned pondweed		I

A2. RARE WETLAND PLANTS OF THE SOUTHERN PRAIRIE POTHOLE REGION

TABLE A2.1

ZONE	SPECIES	FAM	IA	MN	SD	DISTRIBUTION (()indicates historic location only)
MF SEM	Alisma gramineum	ALS	SC			IA-Gree MN - Lyon SD-Broo
MF SEM	Sagittaria engelmannia ssp. brevirostrat	ALS				IA- Emme Webs SD - Broo Mood
MF SEM	Sagittaria graminea	ALS				IA - Webs Gree Stor Dick
SEM	Sagittaria rigida	ALS				IA - Dick Cerr Stor Hami MN - Blue LeSe Wase
SM FN	Berula pusilla	API	T			IA - Clay Dick MN - Chip Nico Blue LacQ SD - Deue
SEM	Peltandra virginica	ARA	E			IA - Gree
SM MF	Aster brachyactis	AST				IA - Emme MN - Kand Linc LacQ Blue SD - Broo Deue Mood
MF SM	Aster falcatus ssp. commutatus	AST				IA - (Clay Palo) MN - Chip LacQ
SM FN	Aster junciformis	AST	T		SC	IA- (Emme Clay Wort) Hanc Dick Winn MN- Chip SD- Broo De
WP SM FN	Cirsium muticum	AST	SC			IA - Winn Wort Cerr Hanc MN - Kand McLe Blue Nico
WP	Solidago riddellii	AST			SC	SD - Deue
OW	Megalodonta beckii	AST	E			IA - (Cerr) Dick MN - Kand LeSe
MF SEM	Callitriche heterophylla Pursh	CAL			SC	IA - (Clay Palo) MN - Nico LacQ
MF SEM	Callitriche verna	CAL			SC	IA - Dick Clay Emme MN - Pipe Redw Murr Chip Kand Nico Cott
FN	Lobelia kalmii	CMP			SC	IA - Clay Cerr Dick Emme Palo MN - Jack Yell Nico LeSe Kand Stea SD - Broo Deue
SM	Carex alopecoidea	CYP			SC	SD - Broo
WP SM	Carex aquatilis	CYP				IA - Clay MN - Yell Nico LeSe SD - Deue
WP SM	Carex aurea	CYP				MN-Lyon Mart SD - Broo
SM FN	Carex diandra	CYP			SC	IA - Emme Webs Winn MN - LeSe Wase
WP SM	Carex hallii	CYP	T			MN - Cott
WP SM	Carex projecta	CYP				IA - Emme Webs Stor Dick MN - Wase
FN	Carex sterilis	CYP	T			MN - Kand LeSe
SM	Carex tribuloides	CYP				IA - Clay MN - Free
FN	Dulichium arundinaceum	CYP				IA - Wort Hanc Winn MN - Blue
MF	Eleocharis parvula	CYP	SC	PT		IA - (Clay Palo) MN - Linc
FN	Eleocharis pauciflora	CYP	SC	SC		IA - (Emme) Dick
SM FN	Eleocharis wolfii	CYP	SC	E		IA - (Emme) MN - (Nico)
SM FN	Eriophorum gracile	CYP	T		SC	IA - (Emme Cerr Webs) Hanc Dick MN - Blue LeSe Wase SD - (Broo)
FN	Rhynchospora capillacea	CYP	T	T	SC	IA - Clay Dick Emme (Palo) MN - Blue Jack Yell SD - Deue

TABLE A2.1. *(Continued)*

ZONE	SPECIES	FAM	IA	MN	SD	DISTRIBUTION ( )indicates historic location only
MF	*Scirpus smithii*	CYP	SC			IA - (Cerr) MN - Blue
FN	*Scleria verticillata*	CYP	T	T		IA - Emme Clay MN- Jack Blue
MF	*Elatine triandra*	ELA	SC			IA - Dick MN - Pipe LacQ Redw Nico
SM FN	*Gentianopsis crinita*	GEN				IA - Emme MN - Kand Jack
SM FN	*Gentianopsis procera*	GEN	SC			IA - Wort Cerr Stor Dick Clay Palo Emme MN - Jack Nico Yell Kand SD -Broo Deue
DEM	*Myriophyllum heterophyllum*	HAL	SC			IA - Emme (Clay Dick Palo Webs Winn) SD - Broo
DEM OW	*Myriophyllum verticillatum*	HAL	SC			IA -Emme (Palo Webs) MN - Nico Kand Chip SD - Broo
MF SEM	*Hippuris vulgaris*	HIP	SC			IA - (Dick Cerr Clay) MN - Pipe Chip Kand Jack
DEM OW	*Vallisneria americana*	HYD				IA - Dick Clay Emme Palo Buen Winn Hanc Webs Hard MN - Jack LeSe Wase
FN	*Triglochin palustre*	JCG	T	SC		IA - Clay Dick (Palo) Emme Osce MN - Yell Murr Jack Blue SD - Broo Deue
FN	*Juncus alpinoarticulatus*	JUN	SC			IA - Dick SD - Deue
MF	*Juncus bufonis*	JUN	SC			IA - (Clay Palo) Dick
FN SEM	*Utricularia intermedia*	LTB	SC			IA - Clay Emme MN - Kand Nico LeSe
FN	*Utricularia minor*	LTB	SC			IA - Dick Emme Hanc MN - Yell
MF	*Rotala ramosior*	LYT	SC	PT		IA - Palo Stor MN LacQ
SM FN	*Menyanthes trifoliata*	MNY	T		SC	IA- (Palo Emme Clay Winn Wort Hanc Stor Hami Webs) Cerr Wrig Dick MN -McLe Blue Wase SD - (Broo)
DEM OW	*Najas marina*	NAJ		SC		MN - Kand
DEM OW	*Brasenia schreberi*	NYM	E			IA - Hanc Wort (Hami) MN - LeSe
WP SM FN	*Cypripedium candidum*	ORC	SC	SC	SC	IA - (Wrig Palo Hami Stor Webs Hard) Koss Clay Dick Cerr Winn SD - Broo MN - Kand Swif Chip Renv McLe Meek LacQ Yell Linc Pipe Redw Nico Blue Wase Free
WP SM	*Cypripedium reginae*	ORC	T			IA- (Emme Winn Webs Hard Stor) Hami MN - Kand Nico Wase
SM FN	*Liparis loeselii*	ORC				IA - Dick Emme Palo Hanc Cerr MN - Yell Blue
FN	*Platanthera hyperborea*	ORC	T			IA - (Stor) Dick Emme Clay MN - Yell Jack Kand Brow Blue Nic
WP SM	*Platanthera praeclara*	ORC	T	E		IA Dick Emme Palo Buen Webs Stor Hami Wrig Hanc Winn Wort Cerr Hard Clay Koss MN - Kand Nico Nobl SD - Broo
WP SM	*Spiranthes cernua*	ORC		SC		IA - Palo Stor MN - LacQ Swif Renv Cott

TABLE A2.1. *(Continued)*

ZONE	SPECIES	FAM	IA	MN	SD	DISTRIBUTION () indicates historic location only
WP SM	*Spiranthes magnicamporum*	ORC	SC		SC	
FN	*Spiranthes romanzoffiana*	ORC	SC			IA - Dick Emme SD - Broo
SM	*Panicum gattingeri*	POA	SC			IA - Clay Dick
SEM	*Zizania aquatica var. interior*	POA			SC	IA - Dick Palo Emme Clay Hanc Cerr Winn MN - Pipe Brow Swif Kand Sibl Chip SD - Broo Mood
OW	*Potamogeton amplifolius*	POT	SC			IA - (Dick Emme Cerr) Hanc MN - Jack Blue LeSe Wase Cott SD - Broo Deue
OW	*Potamogeton bercholdii*	POT				IA - Dick Palo Clay Hanc Gree MN - Pipe
OW	*Potamogeton epihydrus*	POT	SC			IA - Dick (Winn) Hanc
OW	*Potamogeton filiformis*	POT				MN - Kand
OW	*Potamogeton friesii*	POT				IA - Dick Clay Emme Hami MN - Jack Mart Nico McLe LeSe Wase SD - Broo
OW	*Potamogeton praelongus*	POT	SC			IA - Dick (Emme Cerr) MN - Nico Blue
DEM OW	*Potamogeton spirillus*	POT	SC			IA (Dick Emme Cerr Stor Winn) MN - LacQ Linc Mart
DEM OW	*Potamogeton strictifolius*	POT	SC			MN - Linc IA - (Dick)
DEM OW	*Potamogeton vaginatus*	POT			SC	MN - Kand
OW	*Potamogeton vaseyi*	POT	SC	SC		IA - Wort
SEM	*Ranunculus gmelini*	RAN	SC			IA - (Dick) MN - Blue LeSe
SEM DEM	*Ranunculus circinatus var. subrigidus*	RAN	SC			IA - (Clay) MN - Linc Murr SD - Deue Mood
WP	*Geum macrophyllum*	ROS				MN - Linc
SM FN	*Potentilla anserina*	ROS	E			IA - Dick Emme (Hami Hard Wrig) MN - Linc Lyon Yell Chip Pipe SD - Broo Deue Mood
SM FN	*Potentilla palustris*	ROS				IA - Winn Hanc Emme Clay Cerr MN - Meek Blue Wase Brow
SM SEM	*Galium trifidum*	RUB				IA - Hanc Cerr Dick Emme MN - Kand Nico LacQ SD - Broo De
DEM OW	*Ruppia occidentalis*	RUP	SC			IA - Dick MN - Kand SD - Broo
FN	*Salix candida*	SAL	SC		SC	IA- Cerr (Emme Winn Wort Hanc) MN- Yell Nico Kand SD- De
WO	*Salix lucida*	SAL	T			IA -Dick Clay (Winn Wort Hanc Cerr Poca)
SM FN	*Salix pedicellaris*	SAL	T			IA - (Emme Winn) Wort Hanc Cerr MN - Wase LeSe Blue
WP SM	*Salix serissima*	SAL			SC	MN - LeSe
SM FN	*Parnassia glauca*	SAX			SC	IA - Clay Dick Koss Palo Hanc Emme Cerr MN - Jack Mart Yell Nico LeSe SD - Broo Deue
MF SEM	*Bacopa rotundifolia*	SCR			SC	MN - Pipe Lyon Redw LacQ SD- Broo Mood

TABLE A2.1. (*Continued*)

ZONE	SPECIES	FAM	IA	MN	SD	DISTRIBUTION ()indicates historic location only
				STATE STATUS		
SM	*Chelone glabra*	SCR				IA - Palo Webs Wort Cerr Stor MN - Nico Blue LeSe
MF	*Limosella aquatica*	SCR			SC	MN - LacQ Pipe
FN	*Mimulus glabratus var. fremontii*	SCR		T		IA- Emme Dick Winn SD - Broo Mood
SEM	*Sparganium americanum*	SPG				IA- Emme Palo Stor MN - McLe
FN SM	*Pilea fontana*	URT				IA - Dick MN - Renv Kand Blue SD - Broo

A3. KEY TO THE COMMON PLANTS OF FRESHWATER PRAIRIE POTHOLES

Proper plant identification can often be a significant hurdle in monitoring revegetation. While there are several very good plant guides for the region, these include many nonwetland plants and often rely on many technical terms. This identification guide only includes the wetland plants known to occur in the region and avoids many of the technical terms. The plants are arranged here in a "key," or a series of decision steps, that, when followed in series, end at identification of each plant. Technical terms used are illustrated (figures A3.1 and A3.2) and defined in Appendix A4.

At each step, denoted by a pair of numbers (e.g., 27a and 27b), select the choice that best represents the characteristics of the unknown plant. The destination for the next decision step is located at the end of the description of the choice you have selected (e.g., 54). Go to this pair of characteristics and continue the process. Your final destination will be the correct name for the plant (its family is indicated in parentheses).

This plant key only identifies plants to *genus*. In many cases, there is only one member of a genus within the region. So you will have a specific name for the plant. Some plant genera have more than one representative within the region. Other identification guides will need to be consulted to obtain a specific identification. This last step is not hard, once you know the genus of a plant. However, it often requires familiarity with technical terms. The only complete reference for the region is *The Flora of the Great Plains* (1986). This guide includes complete descriptions that are very helpful for confirming plant identifications.

PLANT KEY A3.1

1a. Plants not rooted, free-floating 2
   2a. Plants with stems; leaves deeply divided into narrow segments 3
      3a. Leaves alternate and with bladders..Bladderwort (LTB)
            Utricularia, 3 species
      3b. Leaves whorled, without bladders...Coontail (CER)
            Ceratophyllum demersum
   2b. Plants without obvious stems, leaves not finally dissected...4
      4a. Leaves two-lobed or repeatedly two-forked...5
         5a. Rather thick leaves, floating on the water surface;
            with purplish scales beneath that may look like short "roots".................................Riccia (RIC)
              Ricciocarpus natans
         5b. Rather thin leaves, in tangled massed, below the water surface;
            lacking purple scales beneath..Riccia (RIC)
              Riccia fluitans
      4b. Leaves variously lobed or forked. 6
         6a. Leaves and roots obvious...7
            7a. One root per plant segment, green underside.....................................Duckweed (LMN)
                Lemna, 2 species
            7b. More than one root per plant segment, red underside..........................Duckweed (LMN)
                Spirodela polyrrhiza
         6b. Plant reduced to tiny particles, the size of large sand grains;
            roots not present ....Watermeal (LMN)
              Wolffia columbiana
1b. Plants rooted in sediment, not free-floating unless dislodged 8
   8a. Leaves floating or submersed..9
      9a. All leaves floating on water's surface; leaves shield-shaped to circular............................10
         10a. Leaves shield-shaped, attached to the stalk by the lower surface, often
            near the center; mature leaves less than four inches across .... Water shield (NYM)
               Brasenia schreberi
         10b. Leaves with petiole attached deep in a basal sinus; mature leaves more
            than four inches across...11
            11a. Leaves round; flowers white...White water lily (NYM)
                Nymphaea tuberosa
            11b. Leaves somewhat oval; flowers yellow ....Yellow water lily (NYM)
                Nuphar luteum
      9b. At least some (and maybe all) leaves submersed 12
         12a. Leaves deeply divided into narrow segments...13
            13a. Leaf divisions pinnate. ....Water milfoil (HAL)
                Myriophyllum, 3 species
            13b. Leaf divisions palmate 14
             14a. Flowers in heads..Water marigold (AST)
                Megalodonta beckii
             14b. Flowers not in heads... Water crowfoot (RAN)
                Ranunculus, 3 aquatic species
         12b. Leaves simple, toothed, or with shallow lobes..15
             15a. Leaves in basal rosette, long and ribbon-like. ....Wild celery (HYD)
                Vallisneria americana
             15b. Leaves along stem, leaves various 16
              16a. Leaves opposite or in whorls...17
               17a. Leaves whorled. 18
                18a. Leaves of two kinds: floating leaves
                  spatulate, submersed linear...Water starwort (CAL)
                   Callitriche, 2 species
                18b. Leaves not of two distinct types...19
                19a. Whorls of 3(4-5) leaves..Waterweed (HYD)
                  Elodea, 2 species
                19b. Whorls of 5(6-13) leaves... Mare's tail (HIP)
                  Hippuris vulgaris

PLANT KEY A3.1 *(Continued)*

17b. Leaves opposite 20
    20a. Leaves generally not crowded, seeds
         are flat, toothed on one side....................................Horned pondweed (ZAN)
                                               Zannichellia palustris
    20b. Leaves often crowded, rounded
         tapered fruits.. Bushy pondweed (NAJ)
                                               Najas, 3 species
16b. Leaves alternate. 21
    21a. Leaves not threadlike and no midvein apparent. ....Water star grass (PNT)
                                               Heteranthera, 2 species
    21b. Leaves threadlike, or, if wider, midvein is obvious........................22
        22a. Leaves with inflated, tube-like stipule and
            fruits held on individual stalks.. Widgeon grass (RUP)
                                            Ruppia occidentalis
        22b. Leaves almost always without inflated
           tube-like stipules (only in *Potamogeton*
           vaginatus) and flowers or fruits in dense
           or open spikes.. Pondweed (POT)
                                          Potamogeton, 19 species
8b. Leaves and stems protruding above water's surface or trailing on mudflats...........................23
23a. Plants lacking flowers and seeds, instead possessing spores 24
    24a. Plants a hollow, jointed tube (may have side appendages) that
         readily breaks into sections; no leaves...Scouring rush (EQU)
                                               Equisetum, 4 species
    24b. Plants not a hollow, jointed tube; with large leaves...25
        25a. Leaves with deep lobes but not compound; spore-bearing
           stalk separate from stalks with leaves....................................... Sensitive fern (PLP)
                                             Onoclea sensibilis
        25b. Leaves once or twice compound; spore-bearing sacs
           on underside of leaves.. Marsh fern (PLP)
                                             Thelypteris palustris
23b. Plants possessing flowers and seeds, not spores..26
    26a. Stems and leaves narrow, flat, cylindric or triangular, flowers
        (or flowering heads) never showy...27
    27a. Flowers born in axils of chaffy scales. 28
        28a. Leaves arranged in 3 vertical rows on stem or appearing to
           be absent. The base of the leaf is a tubular closed sheath.
           The stem is round or triangular and usually solid. 29
            29a. Seeds enclosed in a sac, which is inflated or flattened................Sedge (CYP)
                                             Carex, 40 species
            29b. Seeds not enclosed in a sac...30
               30a. Spikelets flattened, scales of spikelets in two
                 vertical rows 31
                 31a. Flowers in clusters at end of stalks .... Nutsedges (CYP)
                                             Cyperus, 9 species
                 31b. Flowers along the stem, in leaf axils...................Three-way sedge (CYP)
                                             Dulichium arundinaceum
            30b. Spikelets rounded. 32
               32a. Seeds with a persistent beak. 33
                   33a. Spikelets few to many; leaves with
                     blades ....Spike rush (CYP)
                                             Eleocharis, 9 species
                 33b. Spikelet solitary; leaves without
                       blades ....Beak rush (CYP)
                                              Rhynchospora capillacea
               32b. Seeds without a persistent beak 34
                 34a. Seeds subtended by a few, short
                     bristles. .... Bulrush (CYP)
                                            Scirpus, 9 species

PLANT KEY A3.1 (*Continued*)

34b. Seeds subtended by many long,
   cottony bristles. ....Cottongrass (CYP)
                      Eriophorum, 2 species

28b. Leaves in 2 vertical rows, the part of the leaf attached to
the stem (sheath) is usually split down one side (open).
Stems usually hollow and round..35

35a. Two or more florets per spikelet. 36

36a. Spikelets of two distinct types; lower branches
 drooping..Wild rice (POA)
                     Zizania palustris

36b. Spikelets all the same. 37

37a. Spikelets arranged into single or numerous
 spikes 38

38a. Spikes single on each plant .... Quackgrass (POA)
                 Agropyron repens

38b. Spikes numerous on each plant; arranged
 along a main axis ....Bearded sprangletop (POA)
              Leptochloa fascicularis

37b. Spikelets arranged in a panicle 39

39a. Long hairs on small stalks that hold the
 spikelets (rachilla); plants very large .... Reed (POA)
              Phragmites communis

39b. Rachilla not hairy; plants various
 in size. 40

40a. Herbage fragrant (sweet);
 three fused florets per spikelet;
 spikelets short, orbicular......................Sweetgrass (POA)
            Hierochloe odorata

40b. Herbage not sweet-smelling;
 number of florets per spikelet
 various...41

41a. Stems solid.....................................Lovegrass (POA)
           Eragrostis hypnoides

41b. Stems hollow. 42

42a. Callus of lemma not
 bearded. 43

43a. Spikelets in a contracted
 panicle, two kinds of glumes -
 one wide and blunt, the
 other thin; plants erect
 ....Prairie wedge grass (POA)
       Sphenopholis obtusata

43b. Spikelets in an open
 panicle; glumes
 not as above; plants
 erect or trailing...............44

44a. Lemmas with awns .Brome (POA)
      Bromus ciliatus

44b. Lemmas without
 easily visible awns.......Manna grass (POA)
      Glyceria, 3 species

42b. Callus of lemma
 bearded.............................45

45a. Callus hairs straight,
 erect, leaf tips
 do not join to
 form a boat-tip. ....White top (POA)
      Scolochloa festucacea

PLANT KEY A3.1 (*Continued*)

PLANT KEY A3.1 *(Continued)*

60b. Leaves not keeled, flowers in dense or open heads.
Each fruit a capsule surrounded by six prong-like
scales..Rush (JUN)
Juncus, 7 species

26b. Leaves usually wider and /or flowers showy.................................61

61a. Leaves with parallel veins..62

62a. Leaves placed edge to edge, forming a flat fan...............................Blue flag (IRD)
Iris shrevei

62b. Leaves not forming a fan...63

63a. Flowers deep blue or purple, in a spike exserted
from a bladed sheath ....Pickerelweed (PNT)
Pontederia cordata

63b. Flowers not blue or purple, flower arrangements
various. 64

64a. Flowers irregular, the lower petal differing
markedly from the others. Leaves linear to oval. 65

65a. Lower lip forming a large pouch. ....Lady slipper (ORC)
Cypripedium, 2 species

65b. Lower lip does not form a pouch. 66

66a. Lower petal ending in a nectar spur. ....Prairie orchid (ORC)
Platanthera, 2 species

66b. Lower petal not ending in a spur 67

67a. Two leaves in a basal pair......................Twayblade (ORC)
Liparis loeselii

67b. Leaves single or several, basal
or along stem, but never a basal
pair...Ladies tresses (ORC)
Spiranthes, 4 species

64b. Flowers regular, with three or six petals,
all the same..68

68a. Flowers yellow to reddish orange.............................69

69a. Flowers yellow, small (less than 1/2"),
leaves basal..Yellow star grass (LIL)
Hypoxis hirsuta

69b. Flowers orange, larger than 1/2", many
leaves whorled. ....Lily (LIL)
Lilium, 2 species

68b. Flowers pink to white...70

70a. Plants with an onion odor, flowers
arranged in an umbel.......................................Wild garlic (LIL)
Allium

70b. Plants odorless, flowers arranged
variously 71

71a. Flowers greenish white in an
open spike. ....Death camas (LIL)
Zigadenus elegans

71b. Flowers in an umbel or panicle,
petals white. 72

72a. Small flowers (less than 1/3"
across); blades never
arrow-shaped. .... .Water plantain (ALS)
Alisma, 2 species

72b. Fewer larger flowers (greater
than 1/3" across); leaves
arrow-shaped, ovate, or linear. ....Arrowhead (ALS)
Sagittaria, 6 species

61b. Leaves with netted venation...73

73a. Woody plants - trees and shrubs. 74

PLANT KEY A3.1 *(Continued)*

74a. Leaves with 3 to 5 leaflets or lobed..Box elder (ACE)
                                      Acer negundo

74b. Leaves toothed but never lobed or divided
 into leaflets..75

  75a. Leaves lanceolate or narrow; buds with
   a single scale..Willow (SAL)
                              Salix, 12 species

  75b. Leaves deltoid or ovate; buds with
   numerous scales. ....Cottonwood (SAL)
                             Populus deltoides

73b. Plants not woody...76

76a. Flowers in heads...77

77a. Leaves and stems spiny ....Thistle (AST)
                              Cirsium, 4 species

77b. Plants not spiny; although may be
 somewhat prickly 78

  78a. Ray flowers yellow or not evident. 79

   79a. At least lower leaves opposite 80

    80a. Leaves on both sides of stem connected
    to each other,(perfoliate) rough leaves,
    square stems...Cup plant (AST)
                        Silphium perfoliatum

    80b. Leaves not perfoliate 81

     81a. Seeds flat, with 2-4 prongs.
     Lower leaves sometimes pinnately
     dissected .... Beggar's Ticks (AST)
                       Bidens, 6 species

     81b. Seeds only somewhat flattened
     with only two bristles that fall often
     when young. Leaves never dissected.
     ....Sunflower (AST)
                      Helianthus, 4 species

   79b. Leaves alternate along stem 82

    82a. Flowering heads with a central disk
    and circle of ray flowers.........................83

     83a. Disk raised above level
     of ray flowers.....................................84

      84a. Leaves entire, disk dark purple
      to brown, somewhat raised...Black-eyed susan (AST)
                      Rudbeckia hirta

      84b. Leaves very deeply lobed to
      divided, disk grayish, much
      raised. ....Gray coneflower (AST)
                      Ratibida pinnata

     83b. Disk at level of ray flowers or
     even depressed.....................................85

      85a. Tall plants (over 3 feet), with
      large, rough, deeply lobed
      leaves .... Compass plant (AST)
                     Silphium laciniatum

      85b. Plants generally smaller..............86

       86a. Ray petals ending in
       3 lobes; ray flowers not
       reflexed on pedicel.................Sneezeweed (AST)
                    Helenium autumnale

       86b. Ray petals not ending in
       lobes; ray flowers reflexed
       on pedicel. ....Ragwort (AST)
                     Senecio, 4 species

PLANT KEY A3.1 (*Continued*)

    82b. Flowering heads without a clearly
        differentiated central disk and
        circle of ray flowers 87
        87a. Many small flowering heads arranged
            in large clusters not forming burs
            at maturity, leaves linear to
            oblong..Goldenrod (AST)
                                    Solidago, 3 species
        87b. Flowering heads in leaf axils,
            forming burs at maturity; leaves
            very broad, somewhat triangular.......Cocklebur (AST)
                                      Xanthium strumarium
  78b. Ray flowers purple, pink, or white...............................88
    88a. Flowering heads not a circle of ray flowers
        surrounding a disk 89
        89a. Ray flowers white, leaves opposite,
            perfoliate..Boneset (AST)
                                  Eupatorium perfoliatum
        89b. Ray flowers purple, leaves opposite,
            but not perfoliate. 90
            90a. Leaves whorled, heads in broad,
                flat panicles. ....Joe pye weed (AST)
                                    Eupatorium maculatum
            90b. Leaves alternate.................................91
                91a. Flower heads in a long, open
                  spike. .... Blazing star (AST)
                                    Liatris, 3 species
                91b. Flower heads in a terminal flat-topped
                  cluster ....Ironweed (AST)
                                    Vernonia fasciculata
    88b. Flowering heads a radiating circle of ray flowers
        surrounding a central disk; flowers white
        to light pink...92
        92a. Stem ribbed, often very hairy.......................Fireweed (AST)
                                    Erechtites hieracifolia
        92b. Stem not ribbed or very hairy 93
            93a. Fall flowering, ray flowers flat.............94
                94a. Achene with 2-4 bristles.................False aster (AST)
                                    Boltonia asteroides
                94b. Achene with numerous hairs ....Aster (AST)
                                    Aster, 7 species
            93b. Spring-flowering, ray
                flowers threadlike. ....Fleabane (AST)
                                    Erigeron philadelphicus
76b. Flowers not in heads...95
  95a. Leaves alternate 96
    96a. Plants with membraneous sheaths on
        the stems. 97
        97a. Leaves at the base of the plant much larger
            than leaves along the stem. Seed enclosed in a
            reddish-brown sac with three wings ....Dock (PLG)
                                    Rumex, 7 species
        97b. Leaves approximately the same size along
            the stem, flowers white to pink. Fruit a black,
            shiny flat to triangular seed, not in a sac with
            wings ....Smartweed (PLG)
                                  Polygonum, 7 species

PLANT KEY A3.1   (*Continued*)

96b. Plants lacking membraneous sheaths on the
  stems...98
  98a. Flowers with four petals (mustards) 99
    99a. Flowers white, plants hairy...........................Cress (BRA)
                                                                    Cardamine, 2 species
    99b. Flowers yellow, plants not hairy.................Marsh cress (BRA)
                                                                    Rorippa palustris
  98b. Flowers with five petals or petals hard
    to count..100
    100a. Tiny flowers arranged as an umbel...........101
      101a. Flowers yellow...............................Meadow parsnip (API)
                                                                    Zizia, 2 species
      101b. Flowers white.....................................102
        102a. Leaves palmately lobed
          or divided...Cow parsnip (API)
                                                                    Heracleum sphondylium
        102b. Leaves pinnately lobed or
          divided..103
          103a. Leaflets of upper leaves
            irreglarly incised ....Water parsnip (API)
                                                                    Berula pusilla
          103b. Leaflets all regularly
            toothed. 104
              104a. Leaves once pinnate;
              main veins not directed to
              sinuses of teeth...............Water parsnip (API)
                                                                    Sium suave
              104b. Leaves usually bipinnate;
              main veins directed to the
              sinuses of teeth........... .Water hemlock (API)
                                                                    Cicuta, 2 species
    100b. Flowers not in umbels. 105
      105a. Small green flowers, petals not
      readily evident...106
        106a. Leaves with scalloped lobes.... ..Meadow rue (RAN)
                                                                    Thalictrum dasycarpum
        106b. Leaves without lobes, but may
          be toothed 107
          107a. Fruit with five beak-like
            projections on top. ....Stone crop (CRA)
                                                                    Penthorum sedoides
          107b. Fruits without beaks 108
            108a. Sepals and bracts surrounding
            flowers stiff and sharp-pointed.
            ....Pigweed (AMA)
                                                                    Amaranthus, 2 species
            108b. Sepals and bracts surrounding
            flowers soft.............................109
              109a. Flowers on lateral, open
                spikes. ....Goosefoot (CHN)
                                                                    Chenopodium, 3 species
              109b. Flowers sessile in leaf
                axils...........................False loosestrife (ONA)
                                                                    Ludwigia polycarpa
      105b. Flowers showy and colorful.................110
        110a. Regular flowers. 111
          111a. Flowers pale blue........Marsh bellflower (CMP)
                                                                    Campanula aparinoides

PLANT KEY A3.1   *(Continued)*

111b. Flowers yellow or white.........112

  112a. Stems and leaves hairy, leaves 3-7 palmate lobes, white flower, purple and yellow tinged............Rose mallow (MLV)
Hibiscus militaris

  112b. Plants other than above......................................113

    113a. Flowers hypogenous (Buttercups).........................114

      114a. Rather large (about 2") oblong to circular simple leaves with lobes at bases - crenate margins.. ....Marsh marigold (RAN)
Caltha palustris

      114b. Leaves not as above. 115

        115a. Leaves with 3-5 lobes; a few white flowers borne on a long stalk................ .Windflower (RAN)
Anemone canadensis

        115b. Leaves various, if simple then less than 2" across. ....Buttercup (RAN)
Ranunculus, 3 species

    113b. Flowers epigynous or perigynous (Roses). 116

    116a. Principal leaves simple, stems woody, flowers white. .... Meadow sweet (ROS)
Spiraea alba

    116b. Principal leaves compound, flowers yellow. 117

      117a. Basal rosette of leaves present..................Avens (ROS)
Geum

      117b. Basal leaves present, but not as a rosette...........Cinquefoil (ROS)
Potentilla, 5 species

110b. Irregular flowers. 118

  118a. Leaves simple. Tubular flowers with 3 lobes forming an upper lip and 2 lobes on either side of a cleft ....Lobelia (CMP)
Lobelia, 3 species

  118b. Leaves compound. Pea-like flowers. 119

    119a. Plants a trailing or climbing vine...Marsh vetchling (FAB)
Lathyrus palustris

    119b. Plants erect, leaves compound.....................................120

      120a. Three leaflets per leaf..................................Tick clover (FAB)
Desmodium canadense

      120b. Many leaflets per leaf..................................Indigo bush (FAB)
Amorpha fruticosa

PLANT KEY A3.1 *(Continued)*

PLANT KEY A3.1 (*Continued*)

134a. Plants with hooks or spines.
....135
    135a. Plants trailing, plants with hooks,
    leaves (4-6) whorled. ....Bedstraw (RUB)
              Galium, 3 species
    135b. Plants erect; plants
    with stinging hairs. ....Stinging nettle (URT)
              Urtica dioica
134b. Plants lacking hooks or
spines. 136
    136a. Plants with square
    stems; ovary 4-lobed. 137
        137a. Strongly aromatic
        ....Field mint (LAM)
              Mentha arvensis
        137b. Not strongly aromatic
       138
            138a. Showy purple petals,
            irregular flowers; the
            floral tube with a distinct
            cap on top..................Skullcap (LAM)
                Scutellaria, 2 species
            138b. Small white regular
            flowers; floral tube
            without a distinct
            cap ....Bugleweed (LAM)
                Lycopus, 3 species
    136b. Plants with round stems
    or only weakly angled
    (sometimes with ridges);
    ovary not 4-lobed. 139
        139a. Stems with ridges, fruits,
        flowers without stalks 140
        140a. Leaves square at attachment
        to stem. 141
            141a. Leaves round, plant
            trailing. .... Water hyssop (SCR)
                Bacopa rotundifolia
            141b. Leaves several times
            longer than broad; plants
            erect...................Toothcup (LYT)
                Ammania, 2 species
        140b. Leaves narrowing to
        the base.at the attachment
        to stem ....Toothcup (LYT)
              Rotala ramosior
    139b. Stems lacking ridges;
    fruits, flowers stalked.
   142
        142a. Showy purple flowers
        ....Monkey flower (SCR)
              Mimulus ringens
        142b. Flowers small and not
        purple............................143
            143a. Leaves toothed.
            144
                144a. Leaves linear.
                .... Hedge hyssop (SCR)
                  Gratiola neglecta

PLANT KEY A3.1 *(Continued)*

144b. Leaves broad.
...........................Clearweed (URT)
Pilea fontana

143b. Leaves entire
...............................145

145a. Petals 2-cleft,
white.................Stitchwort (CRY)
Stellaria, 2 species

145b. Petals not cleft
.................False pimpernel (SCR)
Lindernia dubia

133b. Fruits not in axils, terminal leafy racemes or
spikes..146

146a. Leaves whorled.................Culver's root (SCR)
Veronicastrum virginicum

146b. Leaves opposite 147

147a. Square stems.................148

148a. Aromatic, white flowers in
small heads............Mountain mint (LAM)
Pycnanthemum virginianum

148b. Not aromatic
.....................................149

149a. Upper lip of flowers
very small or apparently
missing ....Germander (LAM)
Teucrium canadense

149b. Upper lip of flowers
present....................150

150a. Blue flowers, regular, in
narrow long terminal spikes,
blooming in rings around spike.
......................Blue vervain (VRB)
Verbena hastata

150b. Flowers pink to rose,
flowers irregular; flowers in
spikes but not long and narrow
............................151

151a. Only one flower in axil
of each bract.
...............False dragonhead (LAM)
Phystostegia virginiana

151b. Several to many flowers
in axils of each bract
.................Woundwort (LAM)
Stachys palustris

147b. Stems not square...............152

152a. Individual flowers greater than
1/2" long or wide. 153

153a. Flowers deep blue to
purple. 154

154a. 4 petals, fringed
at tips.....................Fringed gentian (GEN)
Gentianopsis, 2 species

154b. 5 petals, not
fringed at tips. ....Gentian (GEN)
Gentiana, 2 species

PLANT KEY A3.1 *(Continued)*

153b. Flowers pink to
cream...............................155
  155a. Regular flowers, 5 petals
  ....Prairie phlox (PLM)
              Phlox pilosa
  155b. Irregular flowers.
 156
    156a. Cream white
    flowers, lobes small.
    ....Turtlehead (SCR)
              Chelone glabra
    156b. Pink flowers,
    lobes prominent. ...Gerardia (SCR)
              Agalinis tenuifolia
152b. Individual flowers small..157
157a. Petals notched.............158
  158a. White petals...........Chickweed (CRY)
              Cerastium nutans
  158b. Pink to purple petals
  .... Willow herb (ONA)
              Epilobium, 3 species
157b. Petals not notched.....159
  159a. Flowers in dense short
  heads..........................Fog fruit (VRB)
              Phyla lanceolata
  159b. Flowers solitary on
  pedicels......................Speedwell (SCR)
              Veronica, 3 species

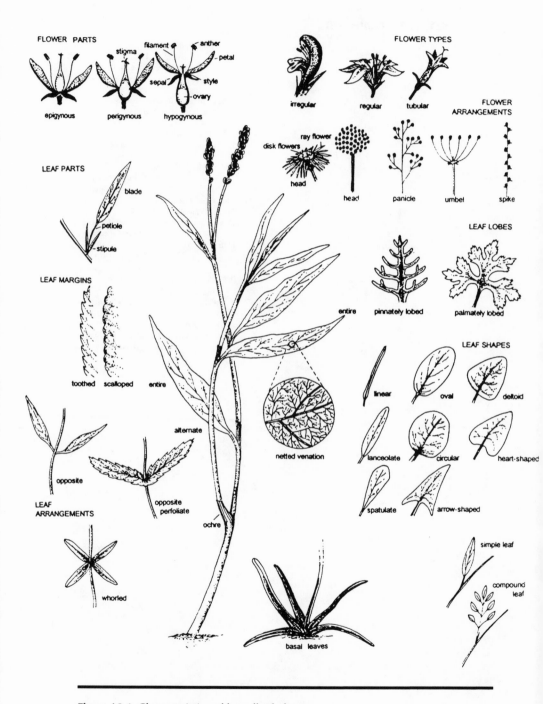

Figure A3.1. Characteristics of broadleaf plants.

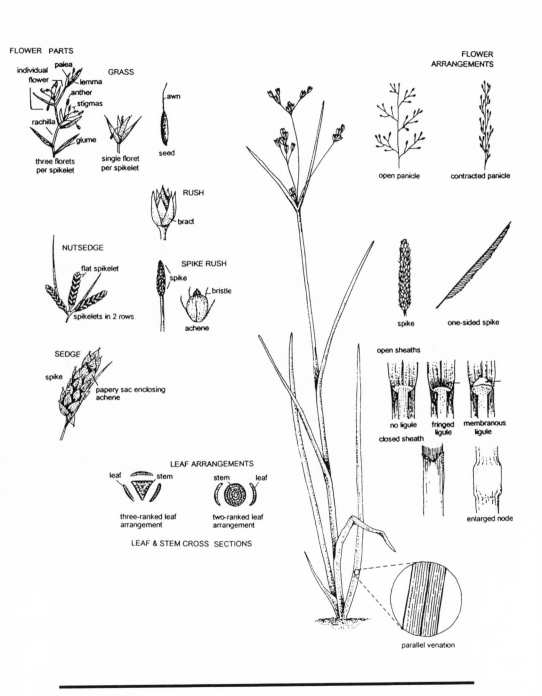

Figure A3.2. Characteristics of grass-like plants.

A4. GLOSSARY OF PLANT IDENTIFICATION TERMS

Grasses, Rushes, Sedges

FLOWERING PARTS

Grass: A floret, or small individual flower, includes male and female reproductive organs and the lemma and palea. Often, several florets will occur above small bracts, called glumes. The glumes and florets, together, are called a spikelet. Glumes or lemmas will sometimes have a needle-like projection, called an awn. Seed may also be awned (needle-like projection). The hard (stiff) parts of the lemma are called callus. The callus of some grass flowers will be hairy or bristly. The small branches between the glumes and lemmas, called rachillas, may also be hairy.

Rush: Rush fruits are a capsule surrounded by six prong-like bracts.

Nutsedges: Nutsedge spikelets are arranged in two rows. These spikelets are flat.

Spike rushes: Spike rush flowers are arranged in a spike at the end of a stem. The flowers and fruits are often hidden from view by papery bracts (small modified leaves). These papery bracts are typical of other members of the sedge family.

Sedges: Sedge fruits are contained within a papery sac that completely envelopes them. The sacs containing the hard fruits are arranged in a spike.

FLOWERING ARRANGEMENTS

Panicle: An irregularly branched flower arrangement. Some panicles will be wide spreading (open panicles), while others will have branches that fold up close to the main stem (contracted panicles).

Spike: Flowers attached directly to the stem, with no side branches supporting flowers or spikelets. Most spikes have flowers positioned around the stem, but a few are one-sided and have all flowers arranged along one edge of the stem.

VEGETATIVE CHARACTERISTICS

Bracts: Reduced, modified leaves.

Leaf venation: The water-conducting pathways, or veins, are obvious on most leaves. The veins of grasses, sedges, and rushes are always arranged parallel to each other.

Sheaths: Sheaths form when leaves are wrapped around the stem for a portion of their length. In some plants, like rushes, sheaths are closed; in grasses sheaths are open.

Nodes: Locations on stems where leaves orginate. Nodes are often somewhat enlarged.

Ligule: Flap of tissue found in grasses at junction of leaf and stem.
Occasionally a fringe of hairs will be present instead of a flap. More
rarely, a ligule will be completely absent.

Leaf arrangement: Leaves will be arranged along the stem in two directions (two-ranked) or in three directions (three-ranked). This can
often be most easily seen by looking straight down on the stem.

Other Flowering Plants

FLOWERING PARTS

Flowers consist of the male organs (anthers and filaments), female
organs (ovary, style, stigma), petals, and other leaf or petal-like structures called sepals. Often, bracts, or other modified leaves, will form a
ring beneath a flower. If the ovary is below the ring where petals emerge,
it is said to be hypogenous; if the ovary is above this ring, it is said to be
epigynous. The flower is perigynous if petals emerge along the bulge of
the ovary.

FLOWER TYPES

Regular flowers: Flowers that are completely symmetrical, or whose
petals radiate out uniformly, are called regular flowers.

Irregular flowers: Those flowers that are bilaterally symmetrical, or that
have two identical halves (like people), are called irregular.

Tubular flowers: Some regular and irregular flowers have petals connected to form a tube, termed tubular flowers.

FLOWER ARRANGEMENTS

Heads: Dense clusters of flowers. Some flowering heads of sunflowers
(and their relatives) include two kinds of flowers and appear almost
as a single flower. The central, short flowers are called disk flowers.
The strap-petalled flowers usually forming an outer ring are ray
flowers.

Panicle: An irregularly branched flower arrangement. Each flower is
supported by a short or long branch.

Umbel: A flat-topped cluster of flowers whose stalks arose from a common point.

Spike: A collection of flowers attached directly to an elongated stem.

VEGETATIVE CHARACTERISTICS

Leaf lobes: When the leaf surface is not at all dissected, it is said to be entire. Leaf surfaces that are dissected, or lobed, may be pinnately or palmately lobed. A pinnately lobed leaf is one whose divisions are arranged along both sides of the main axis of the leaf, where the main vein occurs. A palmately lobed leaf has main veins that radiate from a central place at the base of the leaf. The lobes are oriented along the main routes of these radiating veins.

Simple versus compound leaves: Simple leaves are those that have one continuous surface, even if it is lobed or deeply dissected. In contrast, compound leaves are those whose lobes are actually divisions appearing as several separate leaves all held by the same stalk (petiole).

Stipule: An appendage at the base of a leaf. This appendage may form a tube around the petiole in some pondweeds.

Leaf venation: Leaf venation may be parallel, as in grasses and sedges or netted. Netted venation indicates the veins are not all parallel. In fact, most veins branch across the leaf at various angles to form a network.

Leaf arrangement: Plants that have alternate leaves are those whose leaves are positioned singly at any point on the stem. When pairs of leaves occur at the same point along the stem, they are called opposite leaves. Whorled leaves are those that arise in groups of three or more at any point on the stem.

Leaf shapes: Leaves have various forms, including linear, lanceolate, oval, spatulate, circular or orbicular, deltoid, arrow-shaped, heart-shaped.

Leaf margins: Leaves may have a smooth margin, termed an entire margin. Margins may also be toothed, or serrate, scalloped, or crenate.

Basal leaves: Leaves arising where the stem emerges from the ground. These leaves often form a rosette.

ANIMALS OF THE SOUTHERN PRAIRIE POTHOLE REGION

THIS APPENDIX lists the vertebrate animals that rely on prairie potholes for either nesting or feeding. The second part of the appendix provides distribution information on rare animals of the area. Many other animals may use wetlands for water or occasional cover, but these are not included here. This list was compiled from information included in a number of field guides and references that include the region. These sources are listed here. Tapes of bird calls are available commercially to accompany the Peterson bird guides. Frog-chorusing tapes can be obtained from the Iowa Department of Natural Resources:

> Iowa Department of Natural Resources
> Nongame Program
> Wildlife Research Station
> 1436 255 St.
> Boone, Iowa 50036

Abbreviations Used in Tables

Groups:

AS   Birds with very large area requirements—generally complexes of wetlands and associated grasslands.

OW   Birds that require large semi-permanent wetlands or small lakes that include some open water.

MG Birds that occupy smaller wetlands and require open water with some emergent vegetation.

SB Birds that occupy smaller wetlands, but require well-developed wet prairies, sedge meadows, and shallow emergent vegetation.

DG Dabbling ducks and geese.

SH Birds that nest or forage on bare soil—mud, gravel, or sand—mostly shorebirds.

RA Reptiles and amphibians.

SM Small mammals.

FB Furbearers.

Bird Nest Sites—Based on Weller and Spatcher, 1965:

0 Birds nesting in upland areas, often adjacent to wetlands; sometimes using nesting islands.

1 Birds that select nest sites in marsh edge, low trees, and shrubs.

2 Birds that use short and delicate edge or shallow water emergents such as low sedges and grasses.

3 Birds that prefer tall and robust emergents standing in water, such as cattails and bulrushes.

4 Birds that use low mats of vegetation, often in open areas.

5 Birds that nest on bare ground—mud, gravel, or sand.

Feeding Depths:

FL (Flying)   Birds that feed on flying insects.
VS (Very Shallow) Birds feeding only at the surface of the mud or water.
SH (Shallow)   Birds feeding up to 12 inches into the water.
MD (Moderate)   Birds feeding up to 24 inches into the water.
DP (Deep)   Birds that can feed at depths greater than 24 inches.

Abundance:

R Rare: Of limited abundance and distribution.

Available Information on Prairie Pothole Animals

Bowles, J.B. 1975. *Distribution and Biogeography of Mammals of Iowa.* Special Museum Publication No. 9. Texas Tech University, Lubbock. 184 pp.

Christiansen, J.L. and R.M. Bailey. 1988. *The Lizards and Turtles of Iowa.* Iowa Department of Natural Resources. Nongame Technical Series No. 3. 19 pp.

Christiansen, J.L. and R.M. Bailey. 1990. *The Snakes of Iowa.* Iowa Department of Natural Resources. Nongame Technical Series No. 1. 16 pp.

Christiansen, J.L. and R.M. Bailey. 1991. *The Salamanders and Frogs of Iowa.*

Iowa Department of Natural Resources. Nongame Technical Series No. 3. 24 pp.

Conant, R. and J.T. Collins. 1991. *A Field Guide to Reptiles and Amphibians in Eastern and Central North America*. Third Edition. Peterson Field Guide Series. Houghton Mifflin Company, Boston.

Dinsmore, J.J., T.H. Kent, D. Koenig, P.C. Peterson, and D.M. Roosa. 1984. *Iowa Birds*. Iowa State University Press, Ames.

Hands, H.M., R.D. Drobney, and M.R. Ryan. 1989a. Status of the black tern in the northcentral United States. Report prepared for the U.S. Fish and Wildlife Service, Twin Cities, Minnesota. 15 pp.

Hands, H.M., R.D. Drobney, and M.R. Ryan.1989b. Status of the common loon in the northcentral United States. Report prepared for the U.S. Fish and Wildlife Service, Twin Cities, Minnesota. 26 pp.

Hands, H.M., R.D. Drobney, and M.R. Ryan. 1989c. Status of the least bittern in the northcentral United States. Report prepared for the U.S. Fish and Wildlife Service, Twin Cities, Minnesota. 13 pp.

Hands, H.M., R.D. Drobney, and M.R. Ryan. 1989d. Status of the northern harrier in the northcentral United States. Report prepared by the U.S. Fish and Wildlife Service, Twin Cities, Minnesota. 18 pp.

Harrison, C.V. 1978. *A Field Guide to the Nests, Eggs, and Nestlings of North American Birds*. Collins Publisher, Cleveland, Ohio.

Hazard, E.B. 1982. *The Mammals of Minnesota*. University of Minnesota Press, Minneapolis. 280 pp.

Janssen, R. B. 1987. *Birds in Minnesota*. University of Minnesota Press and James Ford Bell Museum of Natural History, Minneapolis. 352 pp.

Jones, J.K., Jr., D.M. Armstrong, and J.R. Choate. 1985. *Guide to Mammals of the Plains States*. University of Nebraska Press, Lincoln. 371 pp.

Jones, J.K., Jr., D.M. Armstrong, R.S. Hoffman, and C. Jones. 1983. *Mammals of the Northern Great Plains*. University of Nebraska Press, Lincoln. 378 pp.

Roberts, T.S. 1932. *Birds of Minnesota*. University of Minnesota, Minneapolis. Vol. 2.

South Dakota Ornithologists' Union. 1991. *The Birds of South Dakota*. Second Edition. Northern State University Press, Aberdeen, South Dakota. 411 pp.

Weller, M.W. and C.E. Spatcher. 1965. The role of habitat in the distribution and abundance of marsh birds. Iowa State University Agriculture and Home Economics Experiment Station Special Report 43.

196

B1. NESTING BIRDS OF THE SOUTHERN PRAIRIE POTHOLE REGION AND THEIR HABITAT

TABLE B1.1

COMMON NAME	SCIENTIFIC NAME	NEST SITES	FEEDING DEPTH	GROUP	ABUNDANCE
LOONS					
Common loon	*Gavia immer*	3	DP	OW	R
GREBES					
Red-necked grebe	*Podiceps grisegana*	4	DP	OW	R
Horned grebe	*Podiceps auritius*	4	DP	OW	R
Eared grebe	*Podiceps nigricollis*	4	DP	OW	
Western grebe	*Aechmophorus occidentalis*	4	DP	OW	
Pied-billed grebe	*Podilymbus podiceps*	4	DP	OW	
PELICANS					
American white pelican	*Pelecanus erythrorhynchos*	2	SH	OW	R
HERONS & IBISES					
Great egret	*Casnerodius albus*	1	SH	OW	
Great blue heron	*Aredea herodias*	1	SH	OW	
Green-backed heron	*Butorides virescens*	1	SH	OW	
Black-crowned night heron	*Nycticorax nycticorax*	1,3	SH	OW	
Least bittern	*Ixobrychus exilis*	3	SH	SB	R
American bittern	*Botaurus lentiginosus*	2	SH	SB	R
White-faced ibis	*Plegadis chihi*	3	SH	OW	R
DUCKS AND GEESE					
Canada goose	*Branta canadensis*	1	SH-MD	DG	
Trumpeter swan	*Cygnus buccinator*	2	SH	AS	R
Mallard	*Anas platyrhynchos*	0,2	SH	DG	
Gadwall	*Anas strepera*	0	SH-MD	DG	
Northern pintail	*Anas acuta*	0	SH-MD	DG	
Green-winged teal	*Anas crecca*	0,2	SH-MD	DG	
Blue-winged teal	*Anas discors*	0,2	SH-MD	DG	
Northern shoveler	*Anas clypeata*	0,2	SH-MD	DG	
Redhead	*Aythya americana*	3	DP	OW	
Canvasback	*Aythya vallisneria*	3	DP	OW	
Ring-necked duck	*Aythya collaris*	3	DP	OW	
Ruddy duck	*Oxyura jamaicensis*	3	DP	OW	
HAWKS AND OWLS					
Northern harrier	*Circus cyaneus*	2	VS	AS	R
Short-eared owl	*Asio flammeus*	2	VS	AS	R
CRANES AND RAILS					
Whooping crane	*Grus americana*	3	SH	AS	R
Sandhill crane	*Grus canadensis*	2	SH	AS	R
King rail	*Rallus elegans*	2	SH	DG	R
Virginia rail	*Rallus limicola*	2	SH	DG	
Sora	*Porzana carolina*	2	SH	DG	
Common moorhen	*Gallinula chloropus*	3	SH	MG	R
American coot	*Fulica americana*	3	SH	MG	

TABLE B1.1.   (*Continued*)

COMMON NAME	SCIENTIFIC NAME	NEST SITES	FEEDING DEPTH	GROUP	ABUNDANCE
SHOREBIRDS					
Killdeer	*Charadrius vociferus*	5	VS	SH	
Common snipe	*Capella gallinago*	5	VS	SH	
Long-billed curlew	*Numenius americanus*	0	VS	AS	R
Spotted sandpiper	*Actitus macularia*	5	VS	SH	
Willet	*Catoptrophorus semipalmatus*	5	VS	AS	R
Marbled godwit	*Limosa fedoa*	0,2	VS	AS	R
American avocet	*Recurvirostra americana*	2	VS	SH	
Wilson's phalarope	*Phalaropus tricolor*	2	SH	DG	R
GULLS AND TERNS					
Franklin's gull	*Larus pipixcan*	3	VS	OW	R
Forster's tern	*Sterna forsteri*	4	FL	OW	R
Black tern	*Chlidonias niger*	4	VS	MG	R
WRENS					
Sedge wren	*Cistothorus platensis*	2	VS	DG	
Marsh wren	*Cistothorus palustris*	2	VS	DG	
BLACKBIRDS					
Yellow-headed blackbird	*Xanthocephalus xanthocephalus*	3	VS	MG	
Red-winged blackbird	*Agelaius phoeniceus*	3	VS	MG	
SPARROWS					
LeConte's sparrow	*Ammospiza lecontei*	2	VS	DG	R
Savannah sparrow	*Passerculus sandwichensis*	2	VS	DG	
Swamp sparrow	*Melospiza georgiana*	2	VS	DG	
Henslow's sparrow	*Ammodramus henslowii*	2	VS	DG	R
WARBLERS					
Common yellowthroat	*Geothlypis trichas*	2	VS	DG	

B2. MAMMALS OF THE SOUTHERN PRAIRIE POTHOLE REGION

TABLE B2.1

COMMON NAME	SCIENTIFIC NAME	PREFERRED HABITAT	GROUP	ABUND-ANCE
Masked shrew	*Sorex cinereus*	Sedge meadows	SM	
Pygmy shrew	*Microsorex hoyi*	Sedge meadows	SM	R
Short-tailed shrew	*Blarina brevicauda*	Meadows, prairies	SM	
Franklin's ground squirrel	*Spermophilus franklinii*	Wet prairies	SM	R
Meadow vole	*Microtus pennsylvanicus*	Sedge Meadows	SM	
Meadow jumping mouse	*Zapus hudsonius*	Sedge Meadows	SM	
White-footed mouse	*Peromyscus leucopus*	Wet prairies	SM	
Deer mouse	*Peromyscus maniculatus*	Wet prairies	SM	
Muskrat	*Ondatra zibethicus*	Semi-permanent marshes	FB	
Beaver	*Castor canadensis*	Semi-permanent marshes	FB	
Southern bog lemming	*Synaptomys cooperi*	Wet prairies	SM	R
Ermine	*Erminea bangsi*	Meadows, prairies	FB	
Long-tailed weasel	*Mustela frenata*	Meadows, prairies	FB	
Least weasel	*Mustela nivalis*	Meadows, prairies	FB	
Mink	*Mustela vison*	Semi-permanent marshes	FB	
Racoon	*Procyon lotor*	Semi-permanent marshes	FB	

B3. REPTILES AND AMPHIBIANS OF THE SOUTHERN PRAIRIE POTHOLE REGION

TABLE B3.1

COMMON NAME	SCIENTIFIC NAME	PREFERRED HABITAT	GROUP	ABUNDANCE
AMPHIBIANS				
Tiger salamander	*Ambystoma tigrinum*	Variety of wetlands	RA	
Mudpuppy	*Necturus maculosus*	Deep pools	RA	R
Northern leopard frog	*Rana pipiens*	Variety of wetlands	RA	
American toad	*Bufo americanus*	Variety of wetlands	RA	
Canadian toad	*Bufo hemiophrys*	Variety of wetlands	RA	
Chorus frog	*Pseudacris triseriata*	Sedge meadows	RA	
Cricket frog	*Acris crepitans*	Mudflats with open water	RA	R
Bullfrog	*Rana catesbeiana*	Variety of wetlands	RA	Introduced
REPTILES				
Western painted turtle	*Chrysemys picta*	Variety of wetlands	RA	
Snapping turtle	*Chelydra serpentina*	Variety of wetlands	RA	
Blanding's turtle	*Emyoidea blandingi*	Prairie marshes	RA	R
Smooth green snake	*Opheodrys vernalis*	Prairie marshes	RA	R
Western plains garter snake	*Thamnophis radix*	Edge of marshes	RA	
Eastern garter snake	*Thamnophis sirtalis*	Variety of wetlands	RA	

B4. STATUS AND DISTRIBUTION OF RARE ANIMALS IN THE SOUTHERN PRAIRIE POTHOLE REGION

TABLE B4.1

COMMON NAME	SCIENTIFIC NAME	STATUS	CONFIRMED RECORDS (Historic records in ()).
Red-necked grebe	*Podiceps grisegena*	SD: SC	IA: none MN: Nico Lyon (Free Jack) SD: Broo Deue
Horned grebe	*Podiceps auritus*	MN: SC SD: SC	IA: none MN: (Jack McLe Kand) SD: none
Least bittern	*Ixobrychus exilis*	Fed: SC SD:SC	IA: Dick Clay Emme (Gree Hami Palo) MN: Nico SD: Broo Deue Mood
American bittern	*Botaurus lentiginosus*	Fed: SC	IA: Clay Dick (Hanc) MN: (Cott Jack Kand Lyon McLe Meek Nico) SD: Broo Deue Mood
White-faced ibis	*Plegadis chihi*	Fed: C2 SD: SC	IA: Dick MN: Cott Lyon (Jack LacQ) SD: none
American white pelican	*Pelecanus erythrorhynchos*	MN: SC	IA: none MN: LacQ SD: none
Trumpeter swan	*Cygnus buccinator*	X	IA: (Emme Hanc Dick) MN: (Nico Jack Meek) SD: none
Northern harrier	*Circus cyaneus*	IA: End Fed: SC	IA: Cerr Dick Emme Guth Hami Hanc Humb Koss Palo Stor Winn Wrig (Poca) MN: LacQ SD: Deue
Short-eared owl	*Asio flammeus*	IA: End MN:SC	IA: Clay Emme Osce MN: Chip LacQ (Jack LeSe Mart Meek Stea) SD: Deue
Whooping crane	*Grus americana*	IA: X SD: End	IA: (Calh Fran Hanc Winn Wrig Cerr Koss) MN: none SD: (migration sighting: Broo)
Sandhill crane	*Grus canadensis*	IA: X MN: SC	IA: (Dick Hanc Palo Sac) MN: (Jack) SD: none
King rail	*Rallus elegans*	IA: End MN: SC SD: SC	IA: Clay Dick Gree Koss Sac Wrig MN: Blue Cott Free Jack Kand LacQ Lyon Mart Renv SD: Mood
Common moorhen	*Gallinula chloropus*	MN: SC	IA: Clay (Hami) MN: Brow Free Murr Nico Kand Stea (Jack) SD: none
Common loon	*Gavia immer*	IA: X	IA: (Winn) MN: Kand Meek (Jack) SD: none
Long-billed curlew	*Numenius americanus*	X SD: SC	IA: (Buen) MN: none SD: none
Marbled godwit	*Limosa fedoa*	IA: X MN: SC	IA: (Koss) MN: LacQ Brow Swif Yell Kand Stea SD: none
Willet	*Catoptrophorus semipalmatus*		IA: none MN: Jack Lyon Mart (LacQ Swif) SD: none
Wilson's phalarope	*Phalaropus tricolor*		IA: Clay Palo (Hanc) MN: LacQ Linc Nico (Jack McLe) SD: none
Franklin's gull	*Larus pipixcan*		IA: none MN: Jack Kand SD: none

TABLE B4.1. (*Continued*)

COMMON NAME	SCIENTIFIC NAME	STATUS	CONFIRMED RECORDS (Historic records in ()).
Forster's tern	*Sterna forsteri*	IA:SC MN: SC	IA: Cerr Clay Dick Emme Hami Hanc Koss Osce Palo Poca Winn Wort Wrig MN: Jack Kand Lyon Meek Murr Nico Redw SD: none
Black tern	*Chlidonias niger*	IA: SC Fed: SC SD: SC	IA: Carr Cerr Clay Dick Emme Fran Gree Hanc Humb Osce Palo Poca Koss Winn Wort Wrig (Hami) MN: Pipe Linc Lyon Nico LeSe LacQ SD: Deue
Henslow's sparrow	*Ammodramus henslowii*	Fed: SC,C2 IA: Thr MN: SC SD: SC	IA: Cerr Dick Stor (Poca) MN: (Jack LacQ) SD: Broo Deue (Mood)
LeConte's sparrow	*Ammodramus leconteii*	SD: SC	IA: none MN: Linc Lyon SD: none
Blanding's turtle	*Emydoidea blandingi*	Fed: C2 MN:Thr SD: Thr	IA: Clay Gree Hanc Polk Sac (Buen Hami Koss Stor) MN: LeSe Nico Pipe SD: none
Smooth green snake	*Opheodrys vernalis*	SD: SC	IA: Boon Clay MN: SD: none
Cricket frog	*Acris crepitans*	MN: SC	IA: Boon Emme Hard Palo Poca Stor MN: Nobl Pipe SD: none
Mudpuppy	*Necturus maculosus*	IA: X SD: SC	IA: (Hanc) MN: SD: none
Franklin's ground squirrel	*Spermophilus franklinii*		IA: Boon Buen Clay Gree Sac Stor (Cerr Dick Hard Palo Poca Wrig) MN: SD: ?
Southern bog lemming	*Synaptomys cooperi*		IA: Boon Gree (Clay) MN: SD: ?
Pygmy shrew	*Sorex hoyi*	IA: X SD: SC	IA: (Clay) MN: SD: Broo

WETLAND SOILS OF THE SOUTHERN PRAIRIE POTHOLE REGION

IN MOST CASES, modern soil surveys have been published for the counties of the region. Appendix C1 shows the status of soil surveys. Published maps should be verified in the field by comparing the soil on the site to the description in the survey and by using the key provided in Appendix C2. The key groups similar soil series and will identify other series that could be confused with the mapped soil unit.

C1. STATUS OF SOIL SURVEYS (Figure C1.1)

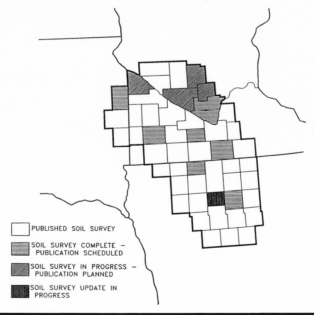

PUBLISHED SOIL SURVEY

SOIL SURVEY COMPLETE – PUBLICATION SCHEDULED

SOIL SURVEY IN PROGRESS – PUBLICATION PLANNED

SOIL SURVEY UPDATE IN PROGRESS

Figure C1.1. Status of soil surveys. (Based on USDA Soil Conservation Service Map # 1000303-07 October 1990.)

C2. HYDRIC SOILS OF THE PRAIRIE POTHOLE REGION

This key includes the wetland soils that occupy upland depressions of southeastern South Dakota, south-central and southwestern Minnesota, and north-central Iowa. The wetlands of this region are prairie potholes that would have been freshwater marshes before agriculture and now primarily are cultivated for corn and soybeans.

This key does not include the remainder of the prairie pothole region northward and westward, which is typified by many alkaline marshes and is currently used primarily for grazing and small grain production. This key also does not include upland wetland soils formed primarily from loess or from Tazewell till, which may occur adjacent to the region covered. Soils of floodplains are also excluded.

BASIC METHOD

Remove a core of soil as deep as possible, but no less than 36 inches in depth. Typically, the soil should be removed in undisturbed sections of 12–18 inches and laid end to end on a tarp or other ground covering. Keep track of which end is from the surface.

A metal plate or dish (muffin pans work well), water, and hydrochloric acid (1M) in a dropper bottle will be required for certain steps of the identification process.

FIGURE C2.1   Soil key.

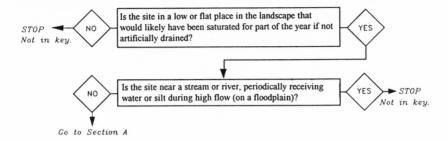

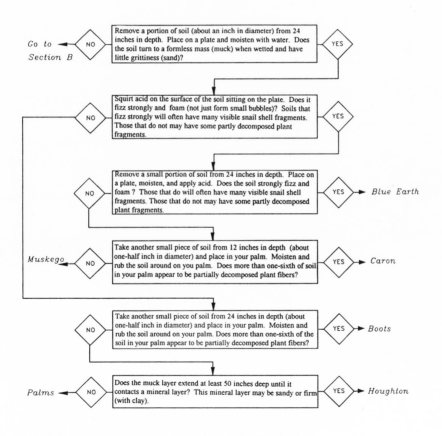

SECTION A

FIGURE C2.1 *(Continued)*

SECTION B

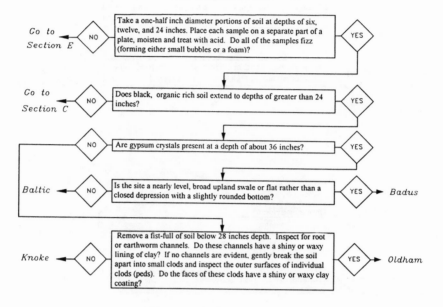

Take a one-half inch diameter portions of soil at depths of six, twelve, and 24 inches. Place each sample on a separate part of a plate, moisten and treat with acid. Do all of the samples fizz (forming either small bubbles or a foam)?
NO → *Go to Section E*
YES

Does black, organic rich soil extend to depths of greater than 24 inches?
NO → *Go to Section C*
YES

Are gypsum crystals present at a depth of about 36 inches?
NO
YES

Is the site a nearly level, broad upland swale or flat rather than a closed depression with a slightly rounded bottom?
NO → *Baltic*
YES → *Badus*

Remove a fist-full of soil below 28 inches depth. Inspect for root or earthworm channels. Do these channels have a shiny or waxy lining of clay? If no channels are evident, gently break the soil apart into small clods and inspect the outer surfaces of individual clods (peds). Do the faces of these clods have a shiny or waxy clay coating?
NO → *Knoke*
YES → *Oldham*

FIGURE C2.1 (*Continued*)

SECTION C

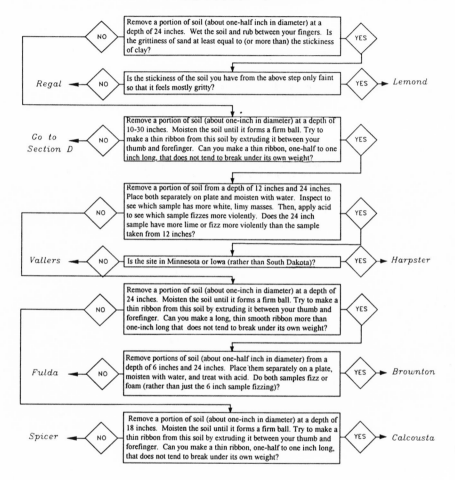

FIGURE C2.1 (*Continued*)

SECTION D

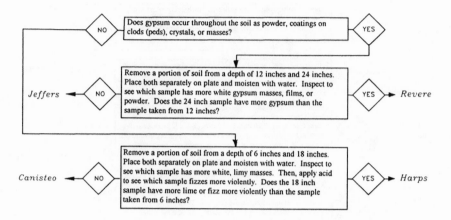

FIGURE C2.1 *(Continued)*

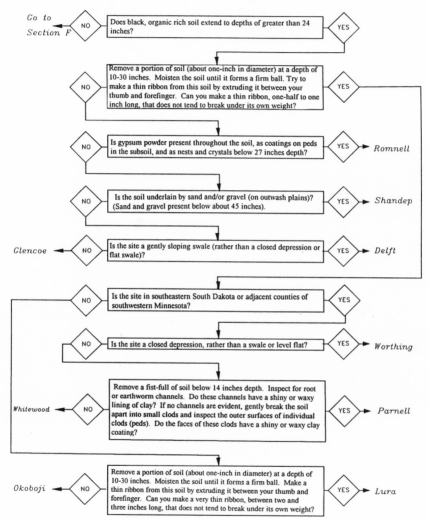

SECTION E

FIGURE C2.1 *(Continued)*

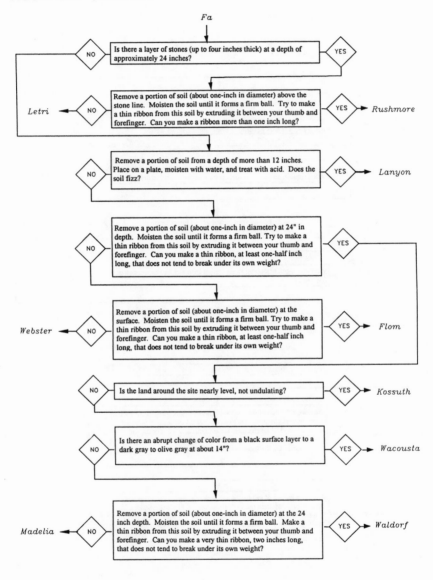

FIGURE C2.1 (*Continued*)

SECTION F

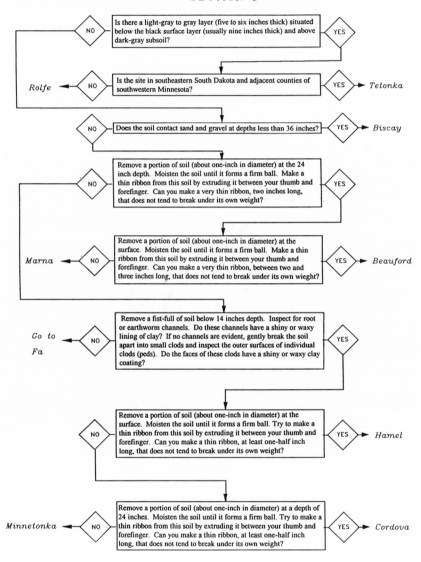

Soil Series Descriptions

Information presented here is, in part, from the National Cooperative Soil Survey. Depths listed here are approximate.

Badus series [fine-silty, mixed (calcareous), mesic Cumulic Haplaquoll] These soils have black silty clay loam surface layers 14 inches thick; the subsoil is black and very dark gray silty clay loam, 20 inches thick. The underlying material is gray and olive-gray silty clay loam. Badus soils are on nearly level, broad upland swales and small flat-bottomed depressions. They formed in calcareous silty alluvium overlying glacial till. An accumulation of gypsum is typical in the B or upper C horizon. Lime segregations are present throughout the solum, decreasing with depth. *Geographic distribution:* East-central and southwestern South Dakota. *Wetland type:* Seasonal prairie potholes.

Baltic series [fine, montmorillonitic (calcareous), mesic Cumulic Haplaquoll] Baltic soils have black silty clay loam surface layers about 11 inches thick; firm black silty clay subsoils with subangular blocky structure; and very dark gray silty clay substrata. The substrata includes many gypsum crystals (at a depth of about 35–45 inches). These soils contain carbonates throughout; snail shells are often common in surface layers. Baltic soils are in flat closed depressions having plane or slightly concave bottoms. They have formed in local clayey alluvial sediments. *Geographic distribution:* Southeastern South Dakota. *Wetland type:* Seasonal prairie potholes.

Beauford series [very fine, montmorillonitic, mesic Typic Haplaquoll] These wet soils are clay throughout. The surface layer is black and very dark gray clay 20 inches thick. The subsoil is mottled olive-gray clay 26 inches thick. The underlying material is grayish brown clay. Beauford soils formed in clayey lacustrine sediments on flats and in slight depressions on lake plains. Loamy glacial till commonly underlies these sediments at depths of less than 10 feet. *Geographic distribution:* South-central Minnesota and possibly north-central Iowa. *Wetland type:* Ephemeral to temporary prairie potholes.

Biscay series [fine-loamy over sandy or sandy-skeletal, mixed, mesic Typic Haplaquoll] Biscay soils developed in outwash and stream alluvium. About 30 inches of loamy material covers sand and gravel. The surface layer, up to 21 inches thick, is black loam grading to very dark gray sandy clay loam. The subsoil is olive-gray loam. Loose loamy and coarse sand occurs below 30 inches. *Geographic distribution:* Central Minnesota, northern Iowa. *Wetland type:* Ephemeral to temporary wet prairie and sedge meadows.

Blue Earth series [fine-silty, mixed (calcareous), mesic Mollic Fluvaquent] These organic matter–rich soils formed in coprogenous earth in drained postglacial lakes. They are in small to large lake basins in glacial moraines and lacustrine plains. These sediments are underlain by loamy glacial till. The surface soil is black and very dark gray mucky silt clay loam 10 inches thick. The substratum is stratified very dark gray and olive-gray mucky silty clay loam. Snail shells are generally present, often abundant. *Geographic distribution:* Southern Minnesota and north-central Iowa. *Wetland type:* Semi-permanent prairie potholes and calcareous fens.

Boots series [euic, mesic Typic Medihemist] Soils formed in organic material more than 51 inches thick within moraines, outwash areas, lake basins, and floodplains. The surface layer is black muck 4 inches thick; the next layer is dark reddish brown muck 6 inches thick. The lower layer is dark reddish brown mucky peat. The organic layer is primarily herbaceous fibers that can still be distinguished when rubbed. *Geographic distribution:* South-central Wisconsin, northern Iowa, southern Minnesota, southern Michigan and New York. *Wetland type:* These soils are primarily in woodlands, but in some places the vegetation is chiefly reeds, sedges, and cattails.

Brownton series [fine, montmorillonitic (calcareous), mesic Typic Haplaquolls] Soils formed in lacustrine sediments over glacial till on glacial lake plains and glacial moraines. The surface soil is black and very dark gray silty clay loam and silty clay 22 inches thick. The subsoil is dark gray silty clay 16 inches thick. The substratum is olive-gray and gray clay loam. Formed in clayey glacial lacustrine sediments or in a mantle of clayey glacial lacustrine sediments and underlying loamy glacial till. These soils have free carbonates in all parts. *Geographic distribution:* South-central Minnesota and possibly north-central Iowa. *Wetland type:* Temporary prairie potholes.

Calcousta series [fine-silty, mixed (calcareous), mesic Typic Haplaquolls] Formed in silty lacustrine sediments in depressions on glacial till plains. The surface soil is black silty clay loam 15 inches thick. The subsoil is olive-gray mottled silty clay loam 6 inches thick. The substratum is olive-gray and light olive-gray mottled silty clay loam. Free carbonates are in all parts of the solum. *Geographic distribution:* Central and north-central Iowa and possibly southern Minnesota. *Wetland type:* Temporary prairie potholes.

Canisteo series [fine-loamy, mixed (calcareous), mesic Typic Haplaquoll] Formed in glacial till under prairie vegetation on rims of depressions on glacial till plains. The surface is black and very dark gray clay loam 20 inches thick. The subsoil is dark gray, dark grayish brown, brown, and olive-gray clay loam 11 inches thick. The substratum is olive-gray clay loam. These soils are on glacial moraines. Typically are calcareous throughout; free carbonates are present between 10 and 20 inches in depth. *Geographic distribution:* Southern Minnesota, northern Iowa, northern Illinois, and eastern South Dakota. *Wetland type:* Ephemeral to temporary prairie potholes and the rims of larger, more permanent basins.

Caron series [coprogenous, euic Limnic Medihemist] Formed in moderately decomposed organic soil material underlain by coprogenous earth under reed and sedge vegetation in bogs. Typically, they have black and very dark brown sapric surface layers 9 inches thick, very dark grayish brown hemic subsoil 37 inches thick, and very dark brown coprogenous earth underlying material. Olive-gray silty clay loam is below the organic material at 84 inches. These soils are in bogs in glacial moraines. The fibers primarily are derived from herbaceous plants, but some pedons contain layers of moss fibers. Snail shells should be abundant; coprogenous earth must be present. Primarily found in bogs in glacial moraines ranging in size from a few hundred acres to several hundred acres. Caron soils typically have a surface layer that is a very dark brown, low-mineral, medium acid to neutral muck. The underlying layer is a brownish low-mineral, medium acid to neutral peaty muck. The next layers are black and very dark brown low-bulk density medium acid to neutral lake sediments. The depth to these sediments ranges from 16 to 51 inches. Loamy materials underlie the lake sediments below depths of 60 inches. *Geographic distribution:* Southern Minnesota and perhaps the southern parts of Wisconsin and Michigan. *Wetland type:* Semi-permanent marshes and perhaps sedge meadows with a prolonged high water table.

Cordova series [fine-loamy, mixed, mesic Typic Argiaquoll] Formed mostly in loamy calcareous glacial till on ground moraines and till plains under mixed hardwoods and grasses in drainageways and in flat uplands. The surface soil is black and very dark gray silty clay loam 14 inches thick. The subsoil is olive-gray silty clay loam in upper 8 inches and dark gray and gray mottled clay loam in lower 17 inches. The substratum is olive mottled clay loam. Many faint dark gray films on faces of peds. *Geographic distribution:* South-central Minnesota and north-central Iowa. *Wetland type:* Wet prairie swales and sedge meadows, sometimes in wooded areas.

Delft series [fine-loamy, mixed, mesic Cumulic Haplaquoll] Formed in loamy alluvium from glacial drift and underlying loamy glacial till on till plains and moraines. Delft soils are found in concave areas on gentle slopes. The surface layer is a black to very dark gray loam to clay loam 29 inches thick. The subsoil is dark gray to olive-gray clay loam or silt loam, 24 inches. The substratum is olive-gray loam. Some pedons have free carbonates in the surface layers, especially in root channels and crayfish burrows. *Geographic distribution:* Southern Minnesota, northern Iowa. *Wetland type:* Wet prairie swales and sedge meadows in drainageways.

Flom series [fine-loamy, mixed, frigid Typic Haplaquolls] Formed in loamy glacial till or glacial lacustrine sediments on moraines. Are found on plane or slightly concave slopes. The surface layer is a very dark gray silty clay loam, silt loam, or clay loam up to 20 inches thick. The subsoil is olive-gray to light olive-gray clay loam. Free carbonates are in parts of this layer in some pedons. The substratum is olive-gray to light olive-gray clay loam or loam. *Geographic distribution:* Western Minnesota and northeastern South Dakota. *Wetland type:* Wet prairie, ephemeral prairie potholes.

Fulda series [calcareous Humic Gley soils] Formed in fine-textured lacustrine material or fine-textured lacustrine material overlying fine-textured till. These soils are found on nearly level topography resembling the smoother ground moraines. Fulda soils are calcareous at/near the surface. The surface layer is a black silt loam to silty clay, 12 to 18 inches deep. The subsurface layer is 15 inches of dark gray-brown to light olive-brown silty clay. The substratum is olive to gray silty clay. The surface is often slightly calcareous. *Geographic distribution:* Southwest Minnesota. *Wetland type:* Wet prairie.

Glencoe series [fine-loamy, mixed, mesic Cumulic Haplaquoll] Formed in loamy sediments from glacial till on glacial moraines. Glencoe soils typically are in closed depressions or low-gradient swales. The surface layer is black clay loam, approximately 35 inches thick. The subsoil is a 13-inch olive-gray loam often with prominent brown mottles. The substratum is grayish brown loam, often with lime segregations. *Geographic distribution:* South-central Minnesota and possibly north-central Iowa. *Wetland type:* Seasonal prairie potholes.

Hamel series [fine-loamy, mixed, mesic Typic Argiaquoll] Formed in local alluvium from glacial till on toe slopes of glacial moraines. Hamel soils occur on concave slopes in swales, rims of closed depressions, and drainageways below steep slopes. The surface layer is a black to very dark gray loam or clay loam, 24 to 30 inches thick. The subsoil is 12–30 inches thick and is dark gray to dark olive-gray loam. Clay films are present on root channels and on ped faces. The substratum is olive-gray loam. *Geographic distribution:* Southeast and south-central Minnesota. *Wetland type:* Wet prairie swales and sedge meadows, sometimes in wooded areas.

Harps series [fine-loamy, mesic Calciaquoll] Soils formed in glacial till or alluvium under prairie vegetation on rims of depressions. Harps soils are on till or outwash plains on narrow rims or shorelines of depressions and on slight rises within poorly defined swales or flat outwash plains. The surface soil is black and very dark gray calcareous clay loam 16 inches thick. The subsoil is light olive-gray, gray, and olive-gray mottled calcareous loam. The substratum is gray, dark gray and dark yellowish brown mottled calcareous loam. *Geographic distribution:* Central and north-central Iowa, south-central Minnesota, and southeastern South Dakota. *Wetland type:* Sedge meadows on rims of seasonal to semi-permanent prairie potholes.

Harpster series [fine-silty, mesic Typic Calciaquoll] Soils formed in silty material derived in calcareous loess or glacial drift. They are on nearly level depressional parts of outwash plains or till plains. The surface soil is black and very dark gray silty clay loam 18 inches thick. The silty clay loam subsoil is dark grayish brown in upper 7 inches, dark gray in next 11 inches, and mixed olive-brown, olive-yellow, and gray in lower 5 inches. The substratum is gray and light olive-brown silt loam and loam. The surface layer is calcareous. These soils commonly contain small snail shells in part or all of the solum. *Geographic distribution:* Central and northern Illinois, east and north-central Iowa, and south-central Minnesota. *Wetland type:* Wet prairie and sedge meadows, temporary prairie potholes.

Houghton series [euic, mesic Typic Medisaprist] Soils formed in herbaceous organic deposits in bogs and other depressional areas within outwash plains, lake plains, till plains, and moraines. The surface layer is black muck 9 inches thick; the underlying layers are black and dark reddish brown sapric material. Organic deposits are more than 51 inches thick. The organic fibers are derived primarily from herbaceous plants. Houghton soils are in bogs ranging from small enclosed depressions to those several hundred acres in size. *Geographic distribution:* Southern Michigan, southern Wisconsin, northern Illinois, northern Ohio, northern Indiana, and northern Iowa. *Wetland type:* Fens, sedge meadows, and shallow wetlands with continuous high water tables or standing water, sometimes associated with lake basins.

Jeffers series [fine-loamy, mixed (calcareous), mesic Typic Haplaquolls] Calcareous soils formed in loamy glacial till or drift on glacial moraines and till plains. The surface layer is 14 to 24 inches thick, black clay loam. The subsoil is dark gray, gray, grayish brown clay loam. The substratum is grayish brown, light olive-brown loam with yellowish brown mottles. Carbonates are disseminated throughout the upper part of the sola and segregated in soft masses in the lower part of the sola. Gypsum occurs throughout the sola in powder form as coats on faces of peds and in nests of crystals. *Geographic distribution:* Southwestern and west-central Minnesota. *Wetland type:* Ephemeral to temporary prairie potholes and the rims of larger, more permanent basins.

Knoke series [fine, montmorillonitic (calcareous), mesic Cumulic Haplaquoll] Formed in calcareous local alluvium in depressions on till plains or in old glacial lakes. The surface layer is black mucky silt loam and silty clay loam, 33 inches thick. The subsoil is black silty clay loam, 13 inches thick. The substratum is gray, dark gray, and very dark gray silty clay loam. Calcareous throughout, often with abundant snail shells. *Geographic distribution:* Central and north-central Iowa and southern Minnesota. *Wetland type:* Seasonal and semi-permanent prairie potholes.

Kossuth series [fine-loamy, mixed, mesic Typic Haplaquoll] Soils formed in moderately fine-textured glacial or lacustrine sediments and in the underlying glacial till on ground moraines. Kossuth soils are typically on plane or concave slopes—the landscape tends to be level or nearly level instead of undulating. Knobs of well-drained soils and depressions are absent or not distinct. The surface layer is black silty clay loam, 20–24 inches thick. The subsoil is very dark gray or dark gray silty clay loam and clay loam. The substratum is olive-gray loam with yellowish brown mottles. *Geographic distribution:* North-central Iowa and possibly south-central Minnesota. *Wetland type:* Wet prairie.

Lanyon series [fine, montmorillonitic, mesic Typic Haplaquoll] Soils that are in broad depressions in the uplands, formed in sediments low in sand content. Formed in silty and clayey lacustrine sediments in depressions on uplands. They have a black light silty clay surface layer about 13 inches thick. The subsoil is very dark gray, dark grayish brown, dark gray, and olive-gray light silty clay, about 7 inches thick. The underlying material is mostly dark gray silty clay and silty clay loam with common mottles. They are calcareous below a depth of about 12 inches. *Geographic distribution:* North-central Iowa and possibly southern Minnesota. *Wetland type:* Temporary prairie potholes.

Lemond series [loam] Lemond soils are found in flat areas and along rims of depressions. These sandy soils were formed in sandy outwash and are calcareous throughout. The top 21 inches is black to very dark gray friable calcareous loam. Friable mottled olive-gray sandy calcareous loam forms a subsoil layer to a depth of 26 inches, below which is loose calcareous sand. *Geographic distribution:* Central Minnesota. *Wetland type:* Wet prairie and sedge meadows.

Letri series [fine-loamy, mixed, mesic Typic Haplaquoll] Formed in glacial till with a thin mantle of loamy erosional sediments or ablation till and underlying firm calcareous loamy glacial till. The upper mantle is lacking in some pedons. These soils are on dissected ground moraine, on plane or slightly concave slopes. Typically, they have black and very dark gray silty clay loam and clay loam surface layers 18 inches thick; dark grayish brown and grayish brown mottled clay loam subsoils 14 inches thick; and grayish brown, mottled loam underlying material. A stone line ranging to as much as 4 inches in thickness is common at the contact of the two sediments. *Geographic distribution:* Southwestern Minnesota and possibly northwestern Iowa. *Wetland type:* Wet prairie.

Lura series [fine, montmorillonitic, mesic Cumulic Haplaquoll] Formed in depressions on clayey glacial lacustrine sediments on glacial lake plains, till plains, and ground moraines. The surface layer is black clay and silty clay approximately 55 inches thick. A subsoil of up to 30 inches thick is present, usually when the surface layer is correspondingly smaller. It is typically silty clay or clay, very dark gray. The substratum is gray silty clay. *Geographic distribution:* South-central Minnesota and possibly north-central Iowa. *Wetland type:* Wet prairies, sedge meadows — temporary prairie potholes.

Madelia series [fine-silty, mixed, mesic Typic Haplaquoll] Formed in lacustrine deposits on lake plains or ground moraines, on plane to slightly concave slopes. The surface soil is black and very dark gray silty clay loam 19 inches thick. The subsoil is olive-gray mottled silty clay loam 8 inches thick. The substratum is olive-gray, olive, and yellowish brown mottled silty clay loam and silt loam. *Geographic distribution:* Southern Minnesota and northern Iowa. *Wetland type:* Wet prairie.

Marna series [fine, montmorillonitic, mesic Typic Haplaquoll] Formed in a mantle of clayey lacustrine sediments underlain by loamy glacial till on lake plains and ground moraines. Soils of slightly concave to slightly convex slopes. Typically, they have black and very dark gray silty clay surface layers 20 inches thick and variegated olive-gray and light olive-brown clay underlying material. *Geographic distribution:* South-central Minnesota and north-central Iowa. *Wetland type:* Wet prairie.

Minnetonka series [fine, montmorillonitic, mesic Typic Argiaquoll] Formed in clayey sediments underlain by loamy, limey glacial till on broad flats and in drainageways in the uplands. The surface layer is black silty clay loam 13 inches thick. The subsoil is very dark gray and olive-gray mottled silty clay and silty clay loam 22 inches thick; clay films are present on peds. The substratum is olive-gray mottled silty clay loam. *Geographic distribution:* Southeastern and south-central Minnesota and northwestern Iowa. *Wetland type:* Wet prairie, sometimes with woodland vegetation.

Muskego series [coprogenous, euic, mesic Limnic Medisaprists] Muskego soils formed in lake basins. These soils formed in organic material overlying coprogenous earth. The surface layer, up to 24 inches thick, is black sapric material with some fibers present when rubbed. The underlying material is black to very dark gray mucky silt loam, with many broken snail shell fragments. *Geographic distribution:* North-central Iowa, south-central Minnesota, central Minnesota. *Wetland type:* Semi-permanent prairie potholes and shallow lake basins.

Okoboji series [fine, montmorillonitic, mesic Cumulic Haplaquoll] Formed in local alluvium under grasses in depressions on till plains. The surface soil is black silty clay loam 32 inches thick. The subsoil is very dark gray, dark gray, and olive-gray silty clay loam 24 inches thick. The substratum is dark gray and gray silty clay loam. *Geographic distribution:* Central and north-central Iowa and southern Minnesota. *Wetland type:* Seasonal wetlands (Okoboji muck — semi-permanent wetlands).

Oldham series [fine, montmorillonitic, calcareous, frigid Cumulic Haplaquoll] Formed in poorly drained soils of wetland depressions. Oldham series are calcareous throughout, with a black to very dark gray silty clay loam surface layer. Fragments of snail shells are common in this layer. The underlying material, below 28 inches, is dark gray calcareous silty clay loam, also with common snail shells. *Geographic distribution:* Eastern South Dakota. *Wetland type:* Seasonal prairie potholes.

Palms series [loamy, mixed, euic, mesic Terric Medisaprist] Formed in deposits of well-decomposed organic material 16 to 50 inches thick over loamy and sandy sediments in depressional areas on lake plains, outwash plains, and hillside seep areas. Sites were formerly lakes or ponds and hillside seep areas. The surface layer is black muck 30 inches thick. The substratum is 15 inches of black and very dark gray clay loam over mottled fine sand. *Geographic distribution:* Southern Michigan, Wisconsin, northern Illinois, Indiana, Iowa, Minnesota, New York, Connecticut, and other northeastern states. *Wetland type:* Semi-permanent wetlands and sedge meadows with prolonged high water table.

Parnell series [fine, montmorillonitic, calcareous, frigid Cumulic Argiaquolls] Soil of depressions. The surface is 34 inches thick, black to very dark gray silty clay loam. The subsoil is very dark gray and olive-gray silty clay loam, about 6 inches thick. The underlying material, found below depths of 40 inches, gray calcareous silty clay loam. *Geographic distribution:* Eastern South Dakota. *Wetland type:* Seasonal prairie potholes.

Regal series [fine-loamy over sandy or sandy-skeletal, mixed (calcareous), mesic Typic Haplaquolls] Calcareous soils formed on broad flats and on swales on outwash plains. A 14-inch thick surface layer of black to very dark gray loam covers a grayish brown mottled calcareous sandy clay loam subsoil. The underlying material is calcareous gravelly loamy sand and gravelly coarse sand. *Geographic distribution:* Central Minnesota. *Wetland type:* Wet prairies and sedge meadows.

Revere series [fine-loamy, mesic Typic Calciaquoll] Formed in calcareous loamy glacial till on ground moraines. Revere soils have slightly convex slopes on rims of depressions and on slight rises within poorly drained flats on ground moraines. The surface layer is about 15 inches thick, black to very dark gray clay loam. The subsoil is dark grayish brown and olive-gray clay loam; substrata of olive-gray loam. Gypsum in powder covers peds and roots' channels at shallow depths; nests of gypsum crystals are common in the subsoil and underlying material. *Geographic distribution:* Southwestern Minnesota. *Wetland type:* Sedge meadows on rims of seasonal to semi-permanent prairie potholes.

Rolfe series [fine, montmorillonitic, mesic Typic Argialboll] Formed under wet prairie vegetation in glacial drift and local alluvium in shallow depressions. The surface layer is black silt loam 9 inches thick. An underlying leached subsoil section is dark gray silt loam 5 inches thick, with common distinct yellowish brown (and sometimes reddish) mottles. Beneath the leached layer is mottled subsoil that is very dark gray and olive-gray silty clay and clay in upper 14 inches and dark gray and olive-gray friable clay loam in lower 17 inches. The substratum is olive-gray mottled clay loam. *Geographic distribution:* North-central Iowa and southern Minnesota. *Wetland type:* Temporary prairie potholes.

Romnell series [fine-loamy, mixed, mesic Cumulic Haplaquoll] Formed in calcareous loamy glacial till or drift on glacial moraines and till plains on broad slightly concave to flat areas. The surface layer is 18 inches thick, black to very dark gray clay loam to silt loam. The subsoil is 25 inches and grades from very dark gray clay loam to olive-gray clay loam to gray clay loam. Clay films are present on peds in the middle portion of the subsoil layer. The substratum is gray clay loam. Gypsum powder coats root channels and ped faces throughout the solum; nests of gypsum crystals are present below 27 inches. This soil is calcareous in the lower part of the subsoil and upper part of substratum. *Geographic distribution:* Southwestern Minnesota. *Wetland type:* Temporary prairie potholes.

Rushmore series [fine-silty, mixed, mesic Typic Haplaquoll] Formed in a mantle of loess or silty lacustrine sediments over calcareous glacial till on ground moraines. On plane or slightly concave slopes . The surface soil is black and dark olive-gray silty clay loam 18 inches thick. The subsoil is olive-gray and olive silty clay loam 10 inches thick. The substratum derived from glacial till is light olive-brown clay loam. A stone line as much as 4 inches thick is common at the contact of the two materials. The soil is calcareous below this contact. *Geographic distribution:* Southwestern Minnesota and possibly northwestern Iowa. *Wetland type:* Wet prairie.

Shandep series [fine-loamy, mixed, mesic Cumulic Haplaquoll] Formed in loam over sand and gravel sediments on stream terraces and outwash plains. The surface soil is black loam and clay loam 29 inches thick. The subsoil is dark gray and gray loam and clay loam 16 inches thick. The substratum is dark gray loamy sand. *Geographic distribution:* Northern and eastern Iowa and southern Minnesota. *Wetland type:* Temporary wetlands.

Spicer series [fine-silty, mixed (calcareous), mesic Typic Haplaquoll] Formed in silty sediments under tallgrasses and sedges on flats and in depressions in silty glacial lacustrine sediments or loess on glacial lake plains and loess-mantled uplands. The surface layer is black and very dark gray silty clay loam 16 inches thick. The subsoil is mottled dark gray and olive-gray silt loam 24 inches thick. The underlying material is mottled olive-gray silt loam. *Geographic distribution:* Southern Minnesota and northern Iowa. *Wetland type:* Wet prairie, sedge meadows, ephemeral and temporary wetlands.

Tetonka series [fine, montmorillonitic, mesic Argiaquic Argialboll] This soil occurs in shallow swales and depressions. Tetonka series formed in these depressions of glacial till and local alluvium. The surface layer, above 9 inches, is a very dark gray and black silt loam with granular structure. Below this, a mottled white to light gray silt loam with platy structure forms a layer 6 inches thick. The subsoil, at depths of 15 to 30 inches is a dark gray silty clay loam with thick coatings. A second subsoil layer, to a depth of 48 inches is light yellowish brown to gray silty clay loam or clay loam. The underlying material is light yellowish brown to gray silt loam, loam, or clay loam with lime segregations. *Geographic distribution:* Eastern South Dakota. *Wetland type:* Wet prairie, sedge meadows, ephemeral and temporary wetlands.

Vallers series [fine-loamy, mixed, frigid Typic Calciaquolls] Soils formed in limy glacial till in drainageways and on rims of depressions. The surface layer is about 17 inches thick and is very limy black or very dark gray silty clay loam. The underlying material is limy mottled grayish clay loam. The surface layer in plowed fields has a grayish cast from dried lime. Fragments of gypsum and snail shells are present in many areas. *Geographic distribution:* Eastern South Dakota *Wetland type:* Sedge meadows on the rims of seasonal and semi-permanent wetlands.

Wacousta series [fine-silty, mixed, mesic Typic Haplaquoll] Formed in silty sediments in depressions or broad swale-like positions on uplands. The surface layer is black silty clay loam 14 inches thick. The subsoil is dark gray and olive-gray mottled silty clay loam 12 inches thick. The substratum is olive-gray and light olive-gray mottled silty clay loam and silt loam. *Geographic distribution:* Central and north-central Iowa and in southern Minnesota and Wisconsin. *Wetland type:* Temporary wetlands.

Waldorf series [fine, montmorillonitic, mesic Typic Haplaquoll] Formed in lacustrine sediments on lake plains or ground moraines. On plane to slightly concave positions. The surface layer is black silty clay loam 20 inches thick. The subsoil is mottled olive-gray silty clay loam and silty clay, 25 inches thick. The underlying material is mottled olive-gray silty clay loam. *Geographic distribution:* Southern Minnesota and northern Iowa. *Wetland type:* Wet prairie, ephemeral prairie potholes.

Webster series [fine-loamy, mixed, mesic Typic Haplaquoll] Formed in glacial sediments and till under prairie vegetation on uplands, nearly plane to slightly concave slopes. The surface soil is black clay loam 17 inches thick. The subsoil is mottled very dark gray, dark gray, and olive-gray clay loam 14 inches thick. The substratum is mottled olive-gray and light olive-gray loam. *Geographic distribution:* North-central and central Iowa and south-central Minnesota. *Wetland type:* Wet prairie, sedge meadows, ephemeral and temporary prairie potholes.

Whitewood series [fine-silty, mixed, mesic Cumulic Haplaquoll] Formed in local silty alluvium on flats and in swales and upland drainageways. These soils have black silty clay loam surface layers 16 inches thick; black, dark gray, and olive-gray silty clay loam subsoils 27 inches thick; and olive-gray silty clay loam underlying material. *Geographic distribution:* Southeastern South Dakota and southwestern Minnesota. *Wetland type:* Wet prairie and sedge meadows, ephemeral and temporary prairie potholes.

Worthing series [fine, montmorillonitic, mesic Typic Argiaquoll] Formed in local clayey alluvium in flat, enclosed depressions. These soils have black silty clay loam surface layers 10 inches thick; mottled black and gray silty clay subsoils 38 inches thick; and mottled gray silty clay loam underlying material. The subsoil shows clay accumulation and often has gypsum nests and manganese concentrations. The presence of carbonates, gypsum, and manganese in the lower subsoil and in substratum varies greatly. *Geographic distribution:* East-central and southeastern South Dakota. *Wetland type:* Wet prairie and sedge meadows, ephemeral and temporary wetlands.

DATA COLLECTION FOR RESTORATION PLANNING AND EVALUATION

D1. RESTORATION PLANNING: BASIN SURVEY INSTRUCTIONS

These instructions are adapted from Wenzel, T. 1992. *Minnesota Wetlands Restoration Guide.* Minnesota Board of Water and Soil Resources. St. Paul, Minnesota.

Equipment needed:

- Instrument (transit, automatic level, magnification 20 times or greater)
- Tripod (wide frame, wooden or aluminum)
- Leveling rod (12 foot minimum)
- Notebook
- Black marking pen (permanent)
- Copy of Agricultural Stabilization and Conservation Service aerial photo (enlarged to fill page)
- Calculator
- Compass accurate to one degree (if not on transit or level)

Procedure:

Most simple surveys can be completed from one instrument location. Inspect the area and select an instrument location that provides a view of the entire basin yet is low enough to be within range of the leveling rod.

A photocopy of the aerial photo can be used to plot reading locations, benchmarks, structure or tile break locations, and the location of the instrument (station). This is also useful as a reference when completing the base map (physical features and limiting factors can be outlines or noted when conducting the survey). Be sure to plot your survey points at the appropriate scale on the aerial photo.

1. The instrument location will be considered the benchmark. Drive a wooden stake that can be relocated to establish this as a temporary benchmark. This location will be arbitrarily assigned an elevation datum of 100 feet.
2. Set the instrument at the selected spot and adjust to level. Establish North as 0 on the surveying instrument.
3. Place the leveling rod at the instrument location and determine the height of the rod at the "eye-end" of the level. This rod-reading (to the nearest 0.1 of a foot) plus 100 feet is recorded as the transit height (TR HT).
4. Move the rod to a point at the intended high-water line. Sight to the rod, center the view with the crosshair, and record the three heights crossing the stadia lines and the direction. Subtract the instrument height from each reading. The middle reading is the elevation of the site. Subtract the bottom reading from the top reading and multiply by 100 for the distance to the rod in feet.
5. Take the necessary shots to run a centerline profile through the basin. Begin downstream of the proposed structure site, in the drainageway, ditch bottom, or tile outlet. Make comments along with the reading to aid in interpreting the data later on.
6. The rod person now can take any intermediate shots or limiting points around the basin. These are points that reflect the ground elevation in the area — not in a hole or on a rock. These readings are recorded the same as the centerline readings. When drainage tile is involved, size of the tile and the elevation of the flowline need to be recorded. Shots for this are taken where the tile line is accessible, usually at an intake.
7. Following intermediate shots, cross sections of any proposed structure sites should be taken. If the structure is to be placed in a ditch, several readings may be needed on both sides of the ditch to determine if the site will allow enough height for a structure.

D2. RESTORATION PLANNING: BASIN SURVEY

SITE_____   DATE_____

(TR HT Transit Height)   All data is calculated in (circle one)   FT   IN   M   CM

POINT	TR HT	TOP STADIA	MIDDLE STADIA	BOTTOM STADIA	BEARING	NOTES	CALCULATED ELEVATION
1							
2							
3							
4							
5							
6							
7							
8							
9							
10							
11							
12							
13							
14							
15							
16							
17							
18							
19							
20							
21							
22							
23							
24							
25							
26							
27							
28							
29							
30							
31							
32							
33							
34							
35							
36							
37							
38							
39							
40							

D3. RESTORATION PLANNING: SITE HISTORY

SITE _____

What year was the marsh drained, or, approximately how many years has the marsh basin been drained?
Have there been any improvements to the drainage system since it was initially drained?
Has the tile or drainage system ever failed? If so, what year(s)? When was the drainage system returned to working order?
How often was there standing water on the land? (Example: more than once a year, once a year, every two years, etc...)
If standing water was present, how long did it take the field to completely drain (Example: 1 day, 2-3 days, a week)
Did you ever notice any wetland plants sprouting in the drained marsh?

D4. RESTORATION PLANNING: PRERESTORATION SITE CONDITION

SITE_____   DATE_____

DRAW A BASE MAP OF THE SITE   + ROADS, PROPERTY BOUNDARIES, CULVERTS, DITCHES, TILE LINES   + PROPOSED STRUCTURES

What soils are on the site - within the basin?

What kinds of marsh plants are present on the site?

Is there evidence of past flooding?

D5. RESTORATION PLANNING: SEEDBANK ASSAY TECHNIQUE

Equipment:

- Sediment-collecting tool (such as long-handled bulb planter, shovel, soil auger)
- Soil sieve of ¼-inch machine cloth (at least 12 inches × 12 inches) attached to a wooden frame or bolted to a heavy plastic bucket with a cutout bottom
- Container with an opening slightly smaller than the soil sieve
- Plastic trays or flats without drain holes—10 per basin (at least 8-inch × 4-inch surface, 2 inches deep)
- Well-lit, protected growing area (such as greenhouse)

Procedure:

1. Collect surface sediment (to a depth of approximately 3 inches) from the basin. A long-handled bulb planter (available at garden centers) works well as a collecting tool in both natural and drained wetlands. Collect samples from around the basin (from at least 20 locations), and mix them together. Since seeds are relatively well mixed across prairie pothole basins, samples from different elevations do not need to be kept separate. Seed bank samples are best collected in early spring (April), when seeds are not actively germinating in the field. If the sediment will not be used immediately, place the sediment in sealed plastic bags and keep cold (40° F).
2. Separate any roots and plant debris from the sediment by passing it through ¼-inch machine cloth. The soil of drained prairie wetlands will often need to be moistened with water and mixed to a thick slurry so it can be sieved.
3. Fill growing trays with sieved sediment to a depth of at least 1.5 inches. Place in a well-lit area protected from rain (pots will wash out during high-intensity rains).
4. Water once or twice daily, as necessary, to maintain saturated soil conditions. Do not water with a high-pressure hose that will dislodge newly germinated seeds from the soil. If seed bank information on submersed aquatics is desired, maintain some flats with at least 1 inch of overlying water.
5. Periodically inspect the seed bank trays for emerging wetland plants. Remove plants when they reach an identifiable stage. Some fast-

growing plants, especially grasses that tiller, will need to be removed before they can be identified so they do not outcompete other seedlings. Carefully remove these plants with their roots and transplant to a separate pot. Maintain seed bank samples for approximately four months to ensure most seeds that can germinate will do so. The few guides available for identifying seedlings focus on agronomic weeds. Often, plants must be grown to maturity to identify them with certainty.

D6. RESTORATION PLANNING: WILDLIFE HABITAT SUITABILITY

SITE _____

♦ SITE CHARACTERISTICS

SIZE OF WETLAND AT FULL POOL (in acres):

CLASS OF WETLANDS (circle):   I   II   III   IV

WHAT PROPORTION OF THE BASIN HAS A DEPTH LESS THAN 6 INCHES AT FULL POOL? _____

WHAT PROPORTION OF THE BASIN HAS A DEPTH LESS THAN 18 INCHES AT FULL POOL? _____

WHAT PROPORTION OF THE BASIN HAS A DEPTH GREATER THAN 36 INCHES AT FULL POOL? _____

⊡ WETLAND COMPLEXES

Record the number and total acreage of wetlands within the nine square miles centered on the section containing the restored wetland:

WETLAND CLASS	APPROXIMATE ACREAGE	NUMBER OF BASINS
CLASS I & II		------------
CLASS III		
CLASS IV		
CLASS V		

○ VEGETATION DEVELOPMENT

For each zone that should be present in the basin (based on wetland class), record the extent of vegetation development:

0 ZONE NOT AT ALL EVIDENT
1 A FEW SCATTERED SEEDLINGS OR ADULT WETLAND PLANTS
2 PATCHES OR NARROW BANDS OF WETLAND VEGETATION
3 CONTINUOUS VEGETATIVE COVER OF WETLAND PLANTS

ZONE	EXTENT	LIST 3 MOST COMMON PLANTS
WET PRAIRIE		
SEDGE MEADOW		
SHALLOW EMERGENT		
DEEP EMERGENT		
OPEN WATER		

■ LAND USE

THE AMOUNT OF LAND IMMEDIATELY ADJACENT TO THE MARSH THAT IS IN PERMANENT COVER : _____

HABITAT SUITABILITY BY GROUP

GROUP	USING INFORMATION FROM ABOVE (symbols coded) IF ANY OF THE FOLLOWING CONDITIONS EXIST THEN....	... THE BASIN IS CURRENTLY SUITABLE. (√)
AS	■ At least 640 acres of adjacent land in permanent cover. ⊡ At least 100 acres of wetland present, including a Class IV basin.	
OW	♦ Class IV or V wetlands. ♦ Wetland size is greater than 40 acres.	
MG	♦ Class III or IV wetlands. ○ Shallow emergent and deep emergent zones with scores of 2.	
SB	○ Wet prairie, sedge meadow, and shallow emergent zones with scores of 3. ♦ Proportion of area less than 18" greater than 50%. ■ A buffer of permanent cover surrounds basin.	
DG	♦ Class II, III, IV wetlands. ♦ Proportion of area less than 18" greater than 50%. ■ A buffer of permanent cover surrounds basin. ⊡ Wetlands of all classes I-IV present in evaluation area.	
SH	○ Sedge meadow and shallow emergent with scores of 0 or 1; Class IV wetlands during drawdown.	
RA	♦ Class II,III,IV wetlands. ■ A buffer of permanent cover surrounds basin	
SM	○ Wet prairie and sedge meadow with scores of 3 ■ A buffer of permanent cover surrounds basin.	
FB	♦ Class IV wetlands. ○ Shallow emergent and deep emergent with scores of 2 or 3.	

D7. RESTORATION DESIGN: CONSTRUCTION WORKSHEET

SITE _____ DATE OF VISIT_____

WHICH OF THE FOLLOWING CONSTRUCTION ACTIVITIES ARE PLANNED FOR THE BASIN:

√	CONSTRUCTION ACTIVITY	DISCUSSED ON PAGES..	
	Grade basin to remove ditch line		
	Plug ditch		See dike calculations below
	Break and remove tile		Length of tile: 1._____ 2._____ (ft) 3._____ 4._____
	Replace removed tile with non-perforated sections		Length of replacement tile required (ft) _____
	Install a pressure release outlet on tile		
	Install a dike to restrict flooding from roads or adjacent property		See dike calculations below
	Install a dike to increase depth of the basin		See dike calculations below
	Install a straight pipe spillway		
	Install a drop inlet spillway		
	Use seepage collar in dike		
	Install Wisconsin Tube		
	Install a fiberglass diaphragm water control structure		
	Outfit outlets with trash racks		
	Cover inlets and outlets with fish barriers		
	Excavate basin to deepen		
	Construct habitat islands		

DIKE SPECIFICATIONS	DESCRIBED ON PAGES...
Depth to clay subsoil (ft): _____ (CH)	
Diaphragm cut-off wall required: YES NO	
Side Slopes of Dike: 1 foot rise for _____ feet width	
Settlement allowance (%): _____	
Dike Dimensions:	Fill Volume Required: Main Dike: $(TW + BW)/2 \times H \times L$ = _____ cu ft. Clay Core: $CW \times CH \times L$ = _____ cu ft Total Fill: Clay Core + Main Dike = _____ cu ft Fll Required = Total Fill x Settlement Allowance + Total Fill

D8. RESTORATION EVALUATION: STRUCTURAL INSPECTION

SITE_____   DATE _____

CONSTRUCTION DATE:_____   DATE OF LAST INSPECTION_____

WATER CONTROL STRUCTURE: Record any problems:
Seepage around structure:
Deterioration (Rust/Rot/Cracking):
Settling:
Erosion:
Sediment reducing capacity of conduit:
Something other than sediment clogging conduit:

EMBANKMENT OF DIKES: Record any problems:
Erosion (Upstream/Downstream):
Seepage:
Sloughing:
Undercutting:
Dike Vegetation:
Rock Rip Rap:
Settlement Apparent:

EMERGENCY SPILLWAY: Record any problems:
Erosion:
Spillway Vegetation:
Rock Rip Rap:
Spillway Last Used:

CONSTRUCTED ISLANDS: Record any problems:
Settling apparent:
Erosion:
Vegetation:

REPAIR OR MAINTENANCE PRIORITY
Priority (P)1. Immediately   2.   This year   3. Next year   4. When Convenient

P	Maintenance or Repair Needed
	Water Control Structure Repairs (see above - seepage, deterioration, erosion)
	Install rack or debris guard on water control structure
	Clean out conduits
	Embankment Repairs (see above - seepage, undercutting, sloughing)
	Embankment Revegetation
	Embankment Improvement (Rip Rap, muskrat screening)
	Emergency Spillway Repairs
	Repairs to Islands

D9. RESTORATION EVALUATION: REVEGETATION

SITE_____

DATE_____

PLANT SPECIES OBSERVED: ◆   ☐

WETLAND CLASS   I   II   III   IV

DATE OF RESTORATION_____

	TALLY THE NUMBER OF SPECIES AND COMPARE TO THESE VALUES:		
Species Group	Depauperate	Typical	Exceptional
WP	0-5	6-14	15 +
SM	0-15	16-30	31 +
SE	0-6	7-15	16 +
DE	0-2	3-4	5 +
SA	0-3	4-10	11 +
FA	0-2	3-5	6

SPECIAL MANAGEMENT CONCERNS:

Y/N

WEEDS:
Are there any invasive weeds on site? Check.
- ☐ Curley leaf pondweed
- ☐ Narrow leaf cattail (or hybrid)
- ☐ Reed canary grass
- ☐ Quackgrass
- ☐ European milfoil
- ☐ Purple loosestrife
- ☐ Canadian thistle

WOODY INVASION:
Is there significant cover of any Group WO species?

LACK OF REVEGETATION:
Is there a "depauperate" rating for any plant group that exists on site (in sites older than 2 yrs)?

PRESENCE OF FEN VEGETATION:
Are at least five Group FN plants present?

RARE PLANTS:
Are there any rare plants present? List Here.

◆ **COVER CATEGORIES:**
R - rare - 1 or few plants, insignificant cover
1 - small groups of plants
2 - approximately 50% cover
3 - nearly complete cover

☐ **GROUPS:**
WP - Wet prairie plants
SM - Sedge meadow plants
SE - Shallow emergent plants
DE - Deep emergent plants
SA - Submersed aquatics
FA - Floating annuals
MF - Mudflat annuals
WO - Woody Plants
FN - Fen Plants
FP - Floating Perennials

D10. RESTORATION EVALUATION: WILDLIFE OBSERVATIONS

SITE

ANIMAL OBSERVED	DATE	SEEN	HEARD	NEST/ YOUNG	TRACKS OR OTHER SIGN	GROUP

SOURCES OF INFORMATION AND SUPPLIES FOR RESTORATIONS

MOST OF THE INFORMATION in the following tables was compiled from the U.S. Soil Conservation Service. 1992. *Directory of Wetland Plant Vendors.* Wetlands Research Program Technical Report WRP-SM-1. U.S. Army Corps of Engineers, Waterways Experiment Station, Vicksburg, Mississippi.

E1. PRAIRIE POTHOLE PLANTS AVAILABLE COMMERCIALLY

This table does not include woody plants, mudflat annuals, and aggressive weeds. For each species, all plant vendors within Iowa, Minnesota, and South Dakota supplying seeds or vegetative propagules are listed. When no plant material sources are available within the region, other sources within the central United States are listed.

TABLE E1.1

AVAILABLE SPECIES	COMMON NAME	PLANT VENDORS (names and addresses provided below)
WET PRAIRIE SPECIES (Group WP)		
Andropogon gerardii	Big bluestem	IA4
Anemone canadensis	Canada anemone	IA4 MN2
Aster simplex	Panicled aster	MN2
Desmodium canadense	Tick clover	IA4
Geum triflorum	Purple avens	IA4
Hypoxis hirsuta	Yellow star grass	MO1
Liatris ligulistylis	Blazing star	IA4
Liatris pycnostachya	Prairie gayfeather	IA4
Lilium michiganense	Michigan lily	IA4
Lobelia spicata	Spiked lobelia	IA4
Panicum virgatum	Switchgrass	MN2
Phlox pilosa	Prairie phlox	IA4
Physostegia virginiana	False dragonhead	IA4 MN1 MN3 MN8 MN9
Ratibida pinnata	Gray coneflower	IA4
Rudbeckia hirta	Black-eyed susan	IA4
Senecio pauperculus	Prairie ragwort	MN2
Silphium perfoliatum	Cup plant	MN2
Thalictrum dasycarpum	Tall meadow rue	IA4 MN2
Veronicastrum virginicum	Culver's root	IA4 MN 2 MN3 MN8
Zigadenus elegans	Death camas	IA4
Zizia aptera	Meadow parsnip	IA4
Zizia aurea	Golden alexander	IA4
SEDGE MEADOW SPECIES (Group SM)		
Amorpha fruticosa	Indigo bush	IA3 MN3 MN8
Agalinis tenuifolia	Gerardia	IA4
Asclepias incarnata	Swamp milkweed	IA4 IA8
Aster novae-angliae	New England aster	IA1 IA4 MN1 MN2 MN3 MN8 MN10
Aster puniceus	Swamp aster	IA4 MN2 MN4
Bidens cernua	Sticktight	MN2
Bidens frondosa	Beggar's ticks	IL1
Boltonia asteroides	False aster	MN1
Bromus ciliatus	Fringed brome	MN2
Calamagrostis canadensis	Bluejoint	MN2
Caltha palustris	Marsh marigold	IA4 MN2
Carex aquatilis	Water sedge	WI5
Carex granularis	Sedge	WI4
Carex retrorsa	Retrorse sedge	WI1 WI7
Carex stricta	Tussock sedge	WI4 WI7
Carex vulpinoidea	Fox sedge	WI4 WI7
Chelone glabra	White turtlehead	IA4 MN2 MN3 MN8
Cirsium muticum	Swamp thistle	MN2
Dulichium arundinaceum	Three-way sedge	IA8
Equisetum hyemale	Common scouring rush	IA8
Eupatorium maculatum	Joe pye weed	IA4 MN2
Eupatorium perfoliatum	Boneset	IA4 MN2
Gentiana andrewsii	Bottle gentian	IA1 IA4 MN2 MN3 MN4 MN7 MN8 MN10
Helenium autumnale	Sneezeweed	MN1 MN2 MN8
Hierochloe odorata	Sweetgrass	MN1
Hypericum majus	St. Johnswort	MN2
Juncus torreyi	Torrey's rush	WI4
Leersia oryzoides	Rice cut grass	WI1 WI2 WI4 WI5 WI7
Lobelia siphilitica	Great lobelia	IA1 IA4 MN1 MN2 MN3 MN6 MN7 MN8
Lycopus americanus	Water horehound	WI2
Lysimachia ciliata	Fringed loosestrife	MN2 MN4
Mimulus ringens	Monkey flower	MN2
Muhlenbergia glomerata	Muhly	MN2

236

TABLE E1.1. (Continued)

AVAILABLE SPECIES	COMMON NAME	PLANT VENDORS
		(names and addresses provided below)
SEDGE MEADOW SPECIES (Group SM) cont.		
Onoclea sensibilis	Sensitive fern	MN4 MN5 MN6 MN7 MN10
Pycnanthemum virginianum	Mountain mint	IA4 MN2
Rumex verticillatus	Water dock	WI4
Scirpus atrovirens	Green bulrush	WI1 WI2 WI3 WI7
Scirpus cyperinus	Wool bulrush	WI1 WI2 WI4 WI7
Senecio aureus	Golden ragwort	MN2
Senecio pauperculus	Prairie ragwort	MN2
Silphium perfoliatum	Cup plant	MN2
Spartina pectinata	Cord grass	IA4 MN2
Stachys palustris	Woundwort	MN2
Teucrium canadense	American germander	IA1 MN6 MN7
Verbena hastata	Blue vervain	IA4 MN2
Vernonia fasciculata	Ironweed	IA1 IA4 MN2 MN3 MN8
EMERGENT SPECIES (Groups SEM DEM)		
Acorus calamus	Sweet flag	IA4 IA8 MN7 MN6 MN8
Alisma triviale	Water plantain	WI1, WI2 WI5, WI7
Carex comosa	Bristly sedge	WI1 WI5
Carex lacustris	Lake sedge	WI1
Carex rostrata	Beaked sedge	WI1 WI2 WI5 WI7
Glyceria striata	Fowl manna grass	WI4
Iris shrevei	Blue flag	IA1 IA4 MN5 MN8
Pontederia cordata	Pickerelweed	IA8
Sagittaria latifolia	Wapato	IA8 MN2
Sagittaria rigida	Arrowhead	WI1 WI5
Scirpus fluviatilis	River bulrush	WI1 WI2 WI5
Scirpus validus	Soft-stemmed bulrush	MN2
Sparganium eurycarpum	Giant burreed	IA4
Typha latifolia	Broad leaf cattail	IA8 MN2
Zizania aquatica	Wild rice	IL1 WI1 WI2 WI4 WI5
SUBMERSED & FLOATING AQUATICS (Groups SA, FA, FP)		
Brasenia schreberi	Water shield	WI1
Lemna minor	Lesser duckweed	IA8
Lemna trisulca	Star duckweed	WI1
Najas flexilis	Bushy pondweed	WI1
Nymphaea tuberosa	White water lily	IL1 WI2 WI5
Potamogeton pectinatus	Sago pondweed	IL1 WI1 WI5 WI6
Potamogeton richardsonii	Richardson's pondweed	WI1 WI2
Ranunculus flabellaris	Yellow water crowfoot	WI1
Ranunculus gmelini	Small yellow crowfoot	WI2
Spirodela polyrhiza	Greater duckweed	IL1 WI1
Vallisneria americana	Wild celery	WI5

E2. PLANT VENDORS

TABLE E2.1

CODE	COMPANY	ADDRESS		PHONE
IA1	Nature's Way	R.R. 1 Box 62	Woodburn, IA 50275	515-342-6246
IA2	Smith Nursery Co.	P.O. Box 515	Charles City, IA 50616	515-228-3239
IA3	Dave's Aquarium and Greenhouse	R.R. 1 Box 97	Kelley, IA 50134	515-769-2446
IA4	Iowa Prairie Seed Co.	Box 228	Sheffield, IA 50475	515-892-4111
IL1	F & J Seed Service	P.O. Box 82	Woodstock, IL 60098	815-338-4029
MN1	Rice Creek Gardens, Inc.	1315 66th Ave. N.E.	Minneapolis, MN 55432	612-574-1197
MN2	Prairie Restorations, Inc.	P.O. Box 327	Princeton, MN 55371	612-389-4342
MN3	Landscape Alternatives, Inc.	1465 N. Pascal St.	St. Paul, MN 55108	612-647-9571
MN4	Orchid Gardens	2232 139th Ave. N.W.	Andover, MN 55304	612-755-0205
MN5	Cooper's Garden	212 W. County Rd. C	Roseville, MN 55113	612-484-7878
MN6	Busse Gardens	Rt. 2 Box 238	Cokato, MN 55321	612-286-2654
MN7	Shady Oaks Nursery	700 19th Ave. N.E.	Waseca, MN 56093	
MN8	Prairie Moon Nursery	Rt. 3 Box 163	Winona, MN 55987	507-452-5231
MN9	Camelot North	R.R. 2 Box 398	Pequot Lakes, MN 56472	218-568-8922
MN10	Ferndale Nursery and Greenhouses	P.O. Box 27	Askov, MN 55074	612-838-3636
MO1	Hi-Mountain Farm	Rt. 2, Box 293	Galena, MN 65656	417-538-4574
WI1	Wildlife Industries, Inc.	P.O. Box 2724	Oshkosh, WI 54903	414-231-3780
WI2	J & J Tranz-Plant	P.O. Box 2532	Oshkosh, WI 54903	414-622-3552
WI3	Branch River Trout Hatchery	8150 River Rd.	Greenleaf, WI 54126	414-864-7761
WI4	Country Wetland Nursery	P.O. Box 126	Muskego WI 53150	414-679-1268
WI5	Kester's Wild Game Food Nurseries, Inc.	P.O. Box 516	Omro, WI 54963	414-685-2929
WI6	LaCrosse Seed Co.	2315 Commerce St.	La Crosse, WI 54602	608-781-4848
WI7	Prairie Ridge Nursery	R.R. 2 9738 Overland Rd.	Mt. Horeb, WI 53572	608-437-5245

E3. SOURCES FOR WATER CONTROL STRUCTURES AND OTHER RESTORATION SUPPLIES

Truax Company
3717 Vera Cruz Ave.
Minneapolis, Minnesota 55422
612-537-6639
(Native grass drills)

Hickenbottom, Inc.
R.R. 2
Fairfield, Iowa 52556
1-800-443-7879 (Outside Iowa)
1-800-247-1723 (In Iowa)
(Inlet/outlet risers)

Fiberglass Utility Supplies, Inc.
R.R. 1 Box 70
Libertyville, Iowa 52567
515-693-3311
(DOS-IR water-level control structure)

Agri-Drain Corporation
1-800-232-4742
(antiseep collars, bar guard intakes)

INDEX

Acorus spp., 138
Acris crepitans, 23
Active revegetation. *See* Revegetation, active
Aechmophorus occidentalis, 22
Agelaius phoeniceus, 6, 21
Agricultural chemicals, and decline in animal species, 30
Agricultural runoff
  nutrients in, 26–27
  and site selection, 43, 44, 46–48
Agricultural Stabilization and Conservation Service, 142, 143
Agropyron repens, 118
Albolls, as wetland restoration sites, 47–48
Algae, controlling growth of, 116
Alisma triviale, 15
Ambystoma tigrinum, 6, 20
American avocet, 22
Ammodramus henslowii, 20
Ammospiza lecontei, 21
Amphibians. *See also* Frogs; Salamanders; Toads
  improving habitat of, 124–25
  minimal habitat requirements, 43, 44
  in restored wetlands, 6–7
  in southern prairie pothole region, 198
  survey of, 105
Anas acuta, 7, 21
Anas clypeata, 20–21
Anas discors, 7
Anas platyrhynchos, 7, 20
Anas strepera, 7, 32
Andropogon geradii, 12
Andropogon scoparius, 12
Animals. *See also* Amphibians; Birds; Fur bearers; Mammals; Reptiles; and specific animals under species or common name
  decline of, in prairie pothole region, 28–31
  distribution of within prairie marshes, 19–23
  groupings of, 193–94
  in southern prairie pothole region, 193–200 (Appendix B)
  status and distribution of rare, 199–200
Antivortex trash racks, 81
Aquolls, 52
Aredea herodias, 22
Arrowhead, 138
Asio flammeus, 20

Assistance programs, 141–42, 144–45
  Conservation Reserve Program, 4, 34, 84, 141, 142–43
  contacts for, 145
  Reinvest in Minnesota, 4, 51, 144
  Wetland Reserve Program, 141, 143–44
Aythya affinis, 32
Aythya americana, 22
Aythya vallisneria, 22

Badger, 33
Basin
  base map, 150
    for evaluating basin, 53–56
  calculating size of, 60–62
  excavation of, 83–84
  extent of, and site selection, 53–56
  hydrology of, 13, 15–16
  methods for increasing water depth in, 83
Basin survey
  equipment for, 222
  form for, 224
  procedure, 222–23
Bear Creek, 95
Bear Lake, 95–96
Beggar's ticks, 15
Bidens cernua, 15
Birds. *See also* Ducks; Geese; Shorebirds; Waterfowl; and specific birds under species or common name
  case study on use of restored wetlands, 109–12
  categories of marsh-breeding, 20
  decline in southern prairie pothole region, 28–30
  feeding depths of, 194
  ground-nesting, 20
  improving habitat for area-sensitive, 120–21
  marsh generalists, 122
  minimal habitat requirements for, 42–43, 44
  molting, 24
  nesting, and their habitat, 195–97
  nesting periods, in southern prairie pothole region, 106
  nest sites, 194
  open water, improving habitat for, 121–22
  in restored wetlands, 6
  secretive, of shallow marshes, 122–23